PERTURBATION METHODS IN HEAT TRANSFER

Series in Computational Methods in Mechanics and Thermal Sciences

W.J. Minkowycz and E.M. Sparrow, *Editors*

Anderson, Tannehill, and Pletcher, Computational Fluid Mechanics and Heat Transfer
Aziz and Na, Perturbation Methods in Heat Transfer
Baker, Finite Element Computational Fluid Mechanics
Shih, Numerical Heat Transfer

PROCEEDINGS
Shih, Editor, Numerical Properties and Methodologies in Heat Transfer: Proceedings of the Second National Symposium

PERTURBATION METHODS IN HEAT TRANSFER

A. Aziz

Professor of Mechanical Engineering
Gonzaga University
Spokane, Washington
Formerly,
University of Riyadh, Saudi Arabia

T. Y. Na

Professor of Mechanical Engineering
University of Michigan, Dearborn

HEMISPHERE PUBLISHING CORPORATION

A subsidiary of Harper & Row, Publishers, Inc.

Washington New York London

DISTRIBUTION OUTSIDE NORTH AMERICA

SPRINGER-VERLAG

Berlin Heidelberg Paris New York Tokyo London

PERTURBATION METHODS IN HEAT TRANSFER

2 3 4 5 6 7 8 9 0 B C B C 8 9 8 7 6

This book was set in Press Roman by Hemisphere Publishing Corporation. The editors were Brenda Brienza and Elizabeth Maggiora; the production supervisor was Miriam Gonzalez; and the typesetter was A. Wayne Hutchins.
BookCrafters, Inc. was printer and binder.

Library of Congress Cataloging in Publication Data

Aziz, A., date
Perturbation methods in heat transfer.

(Series in computational methods in mechanics and thermal sciences)
Bibliography: p.
Includes index.
1. Heat–Transmission–Measurement. 2. Perturbation (Mathematics) I. Na, T. Y. II. Title. III. Series.
QC320.A95 1984 621.402′2 84-6624
ISBN 0-89116-376-X Hemisphere Publishing Corporation
ISSN 0272-4804

DISTRIBUTION OUTSIDE NORTH AMERICA:
ISBN 3-540-13608-8 Springer-Verlag Berlin

To

My parents
My wife, Ayesha
My children, Fahd, Sheza, and Kashif

with love
A. Aziz

CONTENTS

PREFACE

The past two decades have witnessed explosive growth of numerical computation as a tool for heat transfer analysis. The scene is dominated by two powerful procedures: finite differences and finite elements. Notwithstanding this fierce competition, approximate analytical methods have continued to develop and have retained their usefulness. Indeed, some people feel that the combination of analysis and computation may provide a more advantageous method in the future. The recent development of the so-called finite analytic method (Chen et al., 1980) is an example in hand.

A useful approximate analytical tool is the method of perturbation expansion. This subject is covered in several books that are currently available. Of these, books by Bellman (1964), Cole (1968), and O'Malley (1974) are mathematically oriented and not suitable for engineers. The only books with engineering flavor are those of Nayfeh (1973) and Van Dyke (1975a). While the former draws problems from diversified engineering disciplines (except heat transfer), the latter is almost entirely devoted to classical problems in fluid mechanics. Both books are addressed to mature audiences and are certainly too concise and difficult for introductory courses or self-study. Recently Nayfeh (1981) published a more elementary version of his 1973 book.

Despite the steady growth of perturbation literature in heat transfer, no single publication covers this area comprehensively. It was felt that such a coverage would be both timely and useful. The writing of the present book is based on this conviction. This book aims to provide a heat transfer oriented approach to perturbation techniques. The presentation is systematic and sufficiently detailed so that it is easily grasped by a reader with undergraduate background in differential equations, heat transfer, and numerical analysis. At the same time, the development is such that readers are gradually taken to a level where they are exposed to the most recent perturbation literature in heat transfer. The book should therefore serve the dual purpose of a textbook and an up-to-date reference on the subject.

The book is divided into six chapters: Basic Concepts; Regular Perturbation Expansions; Singular Perturbation Expansions; Method of Strained Coordinates; Method of Matched Asymptotic Expansions; and Extension, Analysis, and Improvement of Perturbation Series. Each chapter is profusely illustrated with examples of varying complexity drawn from different areas of heat transfer. A set of problems at the end of each chapter, if fully exploited, should serve to broaden and consolidate the reader's knowledge and skill. The selection of examples is wide enough for the instructor to choose the appropriate material depending upon whether the course is offered at the introductory level or addressed to a more advanced audience.

The book concludes with a bibliography of selected additional heat transfer literature devoted to perturbation analysis. Although we have exercised extreme care, the omission of any significant contribution is simply the result of our unawareness of it. The bibliography enhances the reference value of the book. It should also assist the instructor to identify and formulate additional problems for course work.

Looking at the contents of this book, the most serious omission may appear to be the absence of the method of multiple scales. Despite its current popularity, the method has remained virtually unexploited in heat transfer. Therefore, its inclusion fell beyond the scope of the present book as reflected by its title.

It is usual to acknowledge the contributions of several persons in an undertaking such as this. The senior author owes the greatest debt to his parents whose constant help and guidance have been instrumental in shaping his career, both academic and professional. Next on the list comes his wife, whose understanding and sacrifice has been exemplary. And finally their children, whose tolerance and patience is greatly appreciated.

Jointly, the authors are grateful to Professors W. J. Minkowycz and E. M. Sparrow who encouraged the publication of the book as a title in the Series in Computational Methods in Mechanics and Thermal Sciences. The job of manuscript typing was excellently handled by the senior author's secretary, Mr. Ashraf Choudhry. We deeply appreciate his effort.

A. Aziz
T. Y. Na

CHAPTER

ONE

BASIC CONCEPTS

1.1 INTRODUCTORY REMARKS

Like other branches of engineering, heat transfer analysis during the past two decades has relied heavily on fully numerical procedures involving finite differences and finite elements. Despite this overwhelming trend, approximate analytical methods have continued to develop and provide useful solutions to a variety of problems. One such method is the method of perturbation expansion.

The need to resort to approximation in heat transfer arises due to a number of difficulties. For example, in conduction these difficulties are associated with temperature dependent thermal properties, irregular domain, moving boundary separating the two phases, surface radiation, etc. Similarly, in boundary-layer flows, very limited situations admit similarity solutions in which the governing partial differential equations are reduced to ordinary differential equations. The ones that are of more practical interest are nonsimilar and consequently more difficult to solve. When it comes to radiation, the difficulties are still more formidable because the equations take the form of integro-differential equations.

At least in some cases involving the foregoing difficulties, the mathematical model is amenable to perturbation analysis. To lay the foundation for such analysis, this chapter introduces the basic concepts of perturbation theory.

1.2 PERTURBATION QUANTITY

The key step in any analysis is to translate the physical situation into a mathematical model. To be appropriate for perturbation analysis, the model should be recast into dimensionless form so that the parameters and/or variables governing the system's behavior can be established. Assessing their order of magnitude, one can then identify a parameter or variable that is small compared to others. This is then designated as the perturbation quantity and given the symbol ϵ. To illustrate the procedure of identifying ϵ, we discuss several examples, drawn from different areas of heat transfer. The selection is such that both ordinary and partial differential equations are encountered.

1.2.1 Steady Conduction in a Slab with Variable Thermal Conductivity

As shown in Fig. 1.1, consider one-dimensional conduction in a slab of thickness L made of a material with temperature dependent thermal conductivity k. Assuming that the two faces are maintained at uniform temperatures T_1 and T_2, $T_1 > T_2$, the governing equation and boundary conditions are

$$\frac{d}{dx}\left(k\frac{dT}{dx}\right) = 0 \tag{1.1}$$

$$\begin{aligned} x = 0 \qquad T &= T_1 \\ x = L \qquad T &= T_2 \end{aligned} \tag{1.2}$$

Let the thermal conductivity vary linearly with temperature such that

$$k = k_2[1 + \beta(T - T_2)] \tag{1.3}$$

where k_2 is the thermal conductivity at temperature T_2 and β is a constant. Using Eq. (1.3) and introducing the dimensionless quantities

$$\theta = \frac{T - T_2}{T_1 - T_2} \qquad X = \frac{x}{L} \qquad \epsilon = \beta(T_1 - T_2) = \frac{k_1 - k_2}{k_2} \tag{1.4}$$

into Eqs. (1.1) and (1.2) we obtain

$$\frac{d}{dX}\left[(1 + \epsilon\theta)\frac{d\theta}{dX}\right] = 0 \tag{1.5}$$

$$\begin{aligned} X = 0 \qquad \theta &= 1 \\ X = 1 \qquad \theta &= 0 \end{aligned} \tag{1.6}$$

Here we chose the quantity $\beta(T_1 - T_2)$ or $(k_1 - k_2)/k_2$ as the perturbation quantity ϵ. It is a measure of the variation of thermal conductivity with tem-

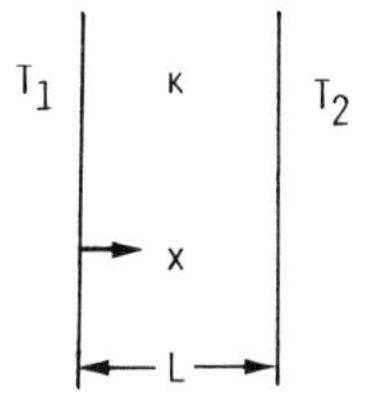

Figure 1.1 Heat conduction in a slab.

perature and may be called the thermal conductivity parameter. To assess the magnitude of ϵ, let us compute ϵ for a pure aluminum slab with $T_1 = 400°C$ and $T_2 = 0°C$. The thermal conductivities are $k_1 = 249$ W/m K and $k_2 = 202$ W/m K giving $\epsilon = 0.23$. By considering a large number of materials and operational temperatures encountered in practice, it is found that $-0.4 < \epsilon < +0.4$ covers most engineering applications (Imber, 1979). Thus, the assumption of small ϵ is justified.

1.2.2 Plane Couette Flow with Variable Viscosity

Consider the steady flow of an incompressible Newtonian fluid between two infinite, parallel plates separated by a distance a as shown in Fig. 1.2. Each plate is maintained at temperature T_0. The lower plate is stationary while the upper plate moves with a uniform velocity V. The thermal conductivity of the fluid is assumed to be constant, but the viscosity is allowed to vary. The pertinent momentum and energy equations are

$$\frac{d}{dy}\left(\mu \frac{du}{dy}\right) = 0 \tag{1.7}$$

$$\frac{d^2T}{dy^2} + \frac{\mu}{k}\left(\frac{du}{dy}\right)^2 = 0 \tag{1.8}$$

$$y = 0 \qquad u = 0 \qquad T = T_0 \tag{1.9}$$

$$y = a \qquad u = V \qquad T = T_0 \tag{1.10}$$

where u = axial velocity
T = temperature
μ = viscosity
k = thermal conductivity

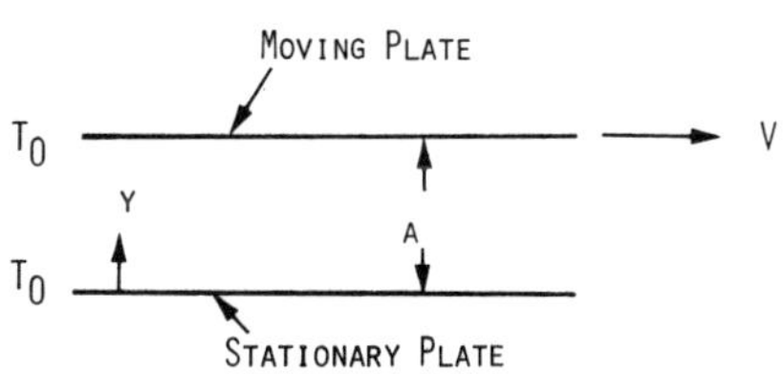

Figure 1.2 Plane couette flow.

Let the viscosity vary exponentially with temperature according to

$$\mu = \mu_0 e^{-\alpha(T-T_0)} \tag{1.11}$$

where μ_0 is the viscosity at T_0 and α is a constant. Introduce Eq. (1.11) and the following dimensionless quantities into Eqs. (1.7)–(1.10)

$$\theta = \frac{T-T_0}{T_0} \qquad Y = \frac{y}{a} \qquad U = \frac{u}{V}$$
$$\beta = \alpha T_0 \qquad \epsilon = \frac{\mu_0 V^2}{kT_0} \tag{1.12}$$

to give (Turian and Bird, 1963)

$$\frac{d}{dY}\left(e^{-\beta\theta}\frac{dU}{dY}\right) = 0 \tag{1.13}$$

$$\frac{d^2\theta}{dY^2} + \epsilon e^{-\beta\theta}\left(\frac{dU}{dY}\right)^2 = 0 \tag{1.14}$$

$$Y = 0 \qquad U = 0 \qquad \theta = 0$$
$$Y = 1 \qquad U = 1 \qquad \theta = 0 \tag{1.15}$$

In this case, the parameter $\mu_0 V^2/kT_0$ is identified as the perturbation quantity ϵ. In viscous flow terminology, it is called the Brinkman number, and represents the ratio of viscous heating to heating due to conduction. Thus, if the effect of viscous heating is weak, one may treat ϵ as small and carry out a perturbation analysis. As shown by Turian and Bird (1963), such an analysis is applicable to flow in a cone-and-plate viscometer.

1.2.3 Laminar Mixed Convection in a Vertical Pipe

Consider the upward flow of a fluid in a uniformly heated vertical pipe of radius a as shown in Fig. 1.3. As the fluid flows, it is heated by combined forced and free convection (mixed convection). The wall temperature T_w increases linearly from $T_w(0)$ at $x = 0$ to $T_w(a)$ at $x = a$. For the fully developed region, Morton (1960) writes the momentum and energy equations as

$$\frac{d^2u}{dr^2} + \frac{1}{r}\frac{du}{dr} = \frac{1}{\nu}\left(\frac{1}{\rho_0}\frac{dp}{dx} + g\right) + \frac{\beta g}{\nu}(T - T_w) \tag{1.16}$$

$$\frac{d^2T}{dr^2} + \frac{1}{r}\frac{dT}{dr} = -\frac{u}{\alpha a}(T_w(a) - T_w(0)) \tag{1.17}$$

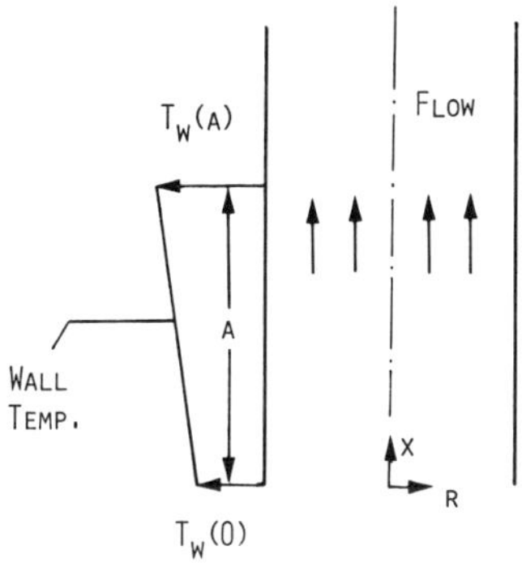

Figure 1.3 Laminar mixed convection in a vertical pipe.

$$r = 0 \qquad \frac{du}{dr} = \frac{dT}{dr} = 0$$
$$r = a \qquad u = 0 \qquad T = T_w \tag{1.18}$$

where u = axial velocity
r = radial coordinate
ν = kinematic viscosity
ρ_0 = density at $T_w(0)$
g = acceleration of gravity
β = coefficient of thermal expansion
T = temperature
α = thermal diffusivity
a = pipe radius

The quantity dp/dx is the pressure gradient for forced flow.

If we introduce the following dimensionless quantities

$$U = \frac{ua}{\alpha} \qquad \theta = \frac{T - T_w}{T_w(a) - T_w(0)} \qquad R = \frac{r}{a}$$
$$P = \frac{a^3(dp/dx + g)}{\alpha\nu\rho_0} \qquad \epsilon = \frac{\beta g a^3 (T_w(a) - T_w(0))}{\alpha\nu} \tag{1.19}$$

Eqs. (1.16)-(1.18) transform to

$$\frac{d^2U}{dR^2} + \frac{1}{R}\frac{dU}{dR} = -P + \epsilon\theta \tag{1.20}$$

$$\frac{d^2\theta}{dR^2} + \frac{1}{R}\frac{d\theta}{dR} = -U \tag{1.21}$$

$$R = 0 \qquad \frac{dU}{dR} = \frac{d\theta}{dR} = 0$$
$$R = 1 \qquad U = \theta = 0 \tag{1.22}$$

The quantity $\beta g a^3 (T_w(a) - T_w(0))/\alpha\nu$ is the familiar Rayleigh number and represents the ratio of buoyancy forces to viscous forces. It can be chosen as the perturbation quantity ϵ if the free convection effect is small compared to forced convection.

1.2.4 Freezing of a Saturated Liquid in Semi-Infinite Region

This example is concerned with a situation where the difficulty is due to the presence of a moving boundary whose position is not known a priori. A relatively simple moving boundary problem is the freezing of a saturated liquid of semi-infinite extent illustrated in Fig. 1.4. Initially the liquid is assumed to be at its freezing temperature T_f. At time $t > 0$, the face at $x = 0$ is maintained at constant subfreezing temperature T_0 so that $T_0 < T_f$. As heat is extracted from the liquid, it begins to freeze. Let the freezing front, at any instant of time, be located at distance x_f. It is assumed that the unfrozen liquid remains at temperature T_f throughout the process. For the solid phase, the applicable equation is that of one-dimensional transient conduction which may be written as

$$\frac{\partial^2 T}{\partial x^2} = \frac{1}{\alpha} \frac{\partial T}{\partial t} \tag{1.23}$$

where α is the thermal diffusivity of the solid phase. The specified temperatures at the wall and at the freezing front give the following conditions

$$T(0, t) = T_0 \qquad T(x_f, t) = T_f \tag{1.24}$$

Since the progress of the freezing front is due to heat conduction, the energy balance at the freezing front gives

$$k \left. \frac{\partial T}{\partial x} \right|_{x=x_f} = \rho L \frac{dx_f}{dt} \tag{1.25}$$

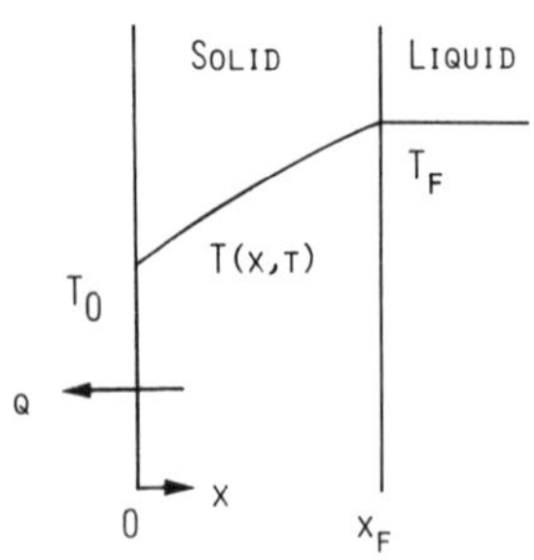

Figure 1.4 Freezing of a saturated liquid in semi-infinite medium.

where k is the thermal conductivity (solid phase), ρ is the density (solid phase), and L is the latent heat.

Let us introduce the dimensionless quantities as

$$\theta = \frac{T_f - T}{T_f - T_0} \qquad X = \frac{x}{x_s} \qquad X_f = \frac{x_f}{x_s}$$
$$\tau = \frac{kt}{\rho c x_s^2} \qquad \epsilon = \frac{c(T_f - T_0)}{L} \tag{1.26}$$

where c is the specific heat (solid phase) and x_s is a reference distance. Next, we change the variables from (X, τ) to (X, X_f). Equations (1.23)-(1.25) are transformed to

$$\frac{\partial^2 \theta}{\partial X^2} = -\epsilon \frac{\partial \theta}{\partial X_f} \frac{\partial \theta}{\partial X}\bigg|_{X=X_f} \tag{1.27}$$

$$\theta(X=0, X_f) = 1 \qquad \theta(X = X_f, X_f) = 0 \tag{1.28}$$

$$\frac{dX_f}{d\tau} = -\epsilon \frac{\partial \theta}{\partial X}\bigg|_{X=X_f} \tag{1.29}$$

Here, our choice of perturbation quantity is the parameter $c(T_f - T_0)/L$ which, in phase change literature, is called the Stefan number. It represents the ratio of the sensible heat to the latent heat stored during the phase change process. The magnitude of Stefan number ϵ can vary considerably depending on the material and the temperature difference involved. For water and paraffin waxes, ϵ is less than unity; for metals it is of the order of 1–10; for materials such as silicates it may reach the order of several hundreds (Solomon, 1979). Thus, for substances such as water and for materials used in latent heat thermal energy storage devices, a perturbation analysis based on the assumption of small ϵ is appropriate.

1.2.5 Two-dimensional Steady Conduction in a Body of Irregular Shape

In practice one often needs to compute temperature distribution in an irregularly shaped body. As an example, consider the problem of two-dimensional steady conduction in the body shown in Fig. 1.5. The eccentricity is denoted by a. The temperature of the inner and the outer surfaces are maintained at T_1 and T_2, respectively. The origin is kept at the center of the inner surface. With respect to the origin, the outer surface can be described by

$$r(\psi) = r_2 + af(\psi) \tag{1.30}$$

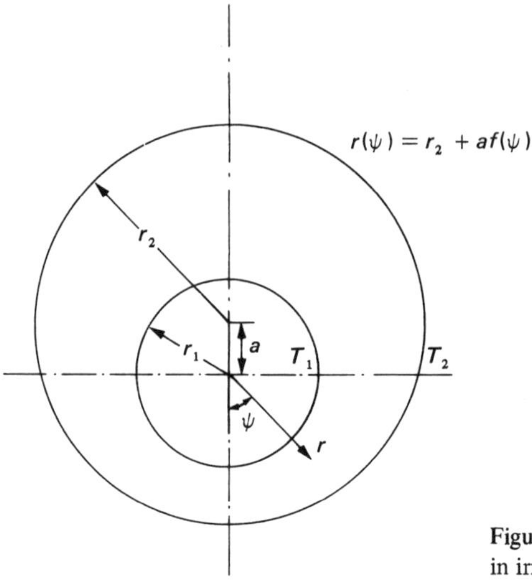

Figure 1.5 Two-dimensional conduction in irregular domain.

where $f(\psi)$ is a prescribable function of ψ. Assuming steady conduction, the equations governing the temperature distribution can be written as

$$\frac{\partial^2 T}{\partial r^2} + \frac{1}{r}\frac{\partial T}{\partial r} + \frac{1}{r^2}\frac{\partial^2 T}{\partial \psi^2} = 0 \tag{1.31}$$

$$\begin{aligned} r &= r_1 \qquad T = T_1 \\ r &= r_2 + af(\psi) \qquad T = T_2 \end{aligned} \tag{1.32}$$

To nondimensionalize Eqs. (1.31) and (1.32) we introduce the following:

$$\begin{aligned} \theta &= \frac{T - T_2}{T_1 - T_2} \qquad R = \frac{r - r_1}{r_2 - r_1} \\ h &= \frac{r_1}{r_2 - r_1} \qquad \epsilon = \frac{a}{r_2 - r_1} \end{aligned} \tag{1.33}$$

and obtain

$$\frac{\partial^2 \theta}{\partial R^2} + \frac{1}{R + h}\frac{\partial \theta}{\partial R} + \frac{1}{(R+h)^2}\frac{\partial^2 \theta}{\partial \psi^2} = 0 \tag{1.34}$$

$$\begin{aligned} R &= 0 \qquad \theta = 1 \\ R &= 1 + \epsilon f(\psi) \qquad \theta = 0 \end{aligned} \tag{1.35}$$

In this formulation the quantity $a/(r_2 - r_1)$ appears as the perturbation quantity. It is a measure of the eccentricity of the inner surface with respect to the outer surface. For concentric surfaces, $\epsilon = 0$, the problem reduces to the familiar one-dimensional conduction in a hollow cylinder. Unlike the previous examples, ϵ now appears in one of the boundary conditions and not in the differential equation. It is apparent that if the eccentricity is slight, one can assume ϵ to be small.

1.2.6 Laminar Natural Convection from a Thin Vertical Cylinder

Except in a few special cases, boundary-layer flows in general are nonsimilar, that is, the governing partial differential equations cannot be reduced to ordinary differential equations. Some of the factors causing nonsimilarities are variable freestream velocity, mass transfer, transverse curvature, and complicated thermal boundary conditions. In this example we consider a situation where the nonsimilarity is due to transverse curvature effect, and show how to identify the appropriate perturbation quantity.

Consider a thin vertical cylinder of radius r_0 maintained at a uniform temperature T_w and convecting naturally to an environment at temperature T_∞ as depicted in Fig. 1.6. Based on the usual Boussinesq model, the continuity, momentum, and energy equations can be written as

$$\frac{\partial}{\partial x}(ru) + \frac{\partial}{\partial r}(rv) = 0 \tag{1.36}$$

$$u\frac{\partial u}{\partial x} + v\frac{\partial u}{\partial r} = g\beta(T - T_\infty) + \frac{\nu}{r}\frac{\partial}{\partial r}\left(r\frac{\partial u}{\partial r}\right) \tag{1.37}$$

$$u\frac{\partial T}{\partial x} + v\frac{\partial T}{\partial r} = \frac{\alpha}{r}\frac{\partial}{\partial r}\left(r\frac{\partial T}{\partial r}\right) \tag{1.38}$$

$$\begin{aligned} r = r_0 \qquad u = v = 0 \qquad T = T_w \\ r = \infty \qquad u = 0 \qquad T = T_\infty \end{aligned} \tag{1.39}$$

where u = axial component of velocity
v = radial component of velocity
g = acceleration of gravity
β = coefficient of thermal expansion
ν = viscosity
α = thermal diffusivity

and where the radial coordinate r is measured from the axis of the cylinder while the axial coordinate x is measured vertically upward such that $x = 0$ corresponds to the leading edge where the boundary layer thickness is zero.

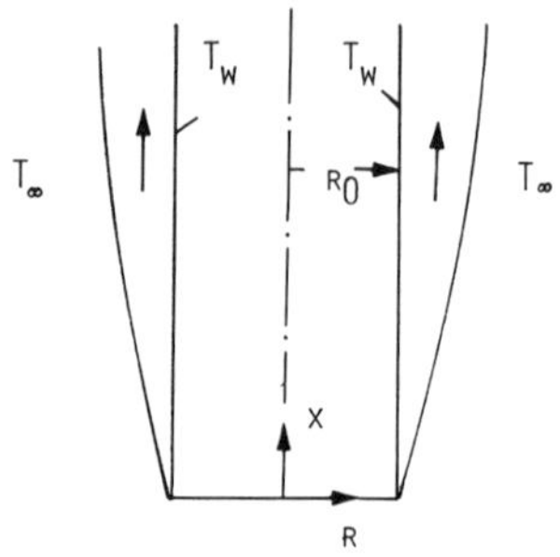

Figure 1.6 Natural convection from a thin vertical cylinder.

By introducing the stream function ψ and the following dimensionless quantities

$$F(\xi, \eta) = \frac{\psi}{4\nu r_0 x^{3/4} \left[g\beta(T_w - T_\infty)/4\nu^2\right]^{1/4}} \tag{1.40}$$

$$\eta = \left[\frac{g\beta(T_w - T_\infty)}{4\nu^2}\right]^{1/4} \frac{r^2 - r_0^2}{2r_0 x^{1/4}} \tag{1.41}$$

$$\theta = \frac{T - T_\infty}{T_w - T_\infty} \tag{1.42}$$

$$\epsilon = \frac{2(x/r_0)^{1/4}}{\left[g\beta(T_w - T_\infty)r_0^3/4\nu^2\right]^{1/4}} \tag{1.43}$$

into Eqs. (1.36)-(1.39), we get (Minkowycz and Sparrow, 1974)

$$(1 + \epsilon\eta)F''' + \epsilon F'' + 3FF'' - 2(F')^2 + \theta = \epsilon\left(F'\frac{\partial F'}{\partial \epsilon} - F''\frac{\partial F}{\partial \epsilon}\right) \tag{1.44}$$

$$(1 + \epsilon\eta)\theta'' + \epsilon\theta' + 3\,\mathrm{Pr}\,F\theta' = \epsilon\,\mathrm{Pr}\left(F'\frac{\partial \theta}{\partial \epsilon} - \theta'\frac{\partial F}{\partial \epsilon}\right) \tag{1.45}$$

$$\begin{aligned} &\eta = 0 \qquad F = F' = 0 \qquad \theta = 1 \\ &\eta = \infty \qquad F' = 0 \qquad \theta = 0 \end{aligned} \tag{1.46}$$

where primes denote differentiation with respect to η and $\mathrm{Pr} = \nu/\alpha$ is the Prandtl number. The perturbation quantity defined by Eq. (1.43) is called the transverse curvature parameter. In the limit, as $r_0 \to \infty$, $\epsilon \to 0$, and the flow situation reduces to that for a vertical plate. Thus, ϵ is a measure of curvature effect and becomes significant for thin cylinders. It is interesting to note that, unlike previous examples, ϵ now represents a coordinate.

1.2.7 Unsteady Heat Transfer for Laminar Flow over a Flat Plate

As an example of unsteady boundary-layer, let us consider a semi-infinite flat plate with an incompressible fluid flowing past it with a uniform velocity U_∞.

It is assumed that initially the plate and the fluid are at the same temperature T_∞, so that heat transfer is absent. We have only the velocity boundary-layer which is described by the well known Blasius equation. At time $t=0$, the plate temperature is suddenly raised to T_0, and we are interested in the development of the thermal boundary-layer. The temperature difference $(T_0 - T_\infty)$ is stipulated to be small enough so as not to disturb the already established velocity field. The appropriate unsteady boundary-layer equations are (Riley, 1963)

$$\frac{\partial u}{\partial x} + \frac{\partial v}{\partial y} = 0 \tag{1.47}$$

$$u\frac{\partial u}{\partial x} + v\frac{\partial u}{\partial y} = \nu\frac{\partial^2 u}{\partial y^2} \tag{1.48}$$

$$\frac{\partial T}{\partial t} + u\frac{\partial T}{\partial x} + v\frac{\partial T}{\partial y} = \alpha\frac{\partial^2 T}{\partial y^2} \tag{1.49}$$

$$\begin{aligned} &y = 0 \quad u(x,0)=0 \quad v(x,0)=0 \quad T(x,0,t)=T_w \\ &y=\infty \quad u(x,\infty)=U_\infty \quad T(x,\infty,t)=T_\infty \\ &t=0 \quad T(x,y,0)=T_\infty \end{aligned} \tag{1.50}$$

where x = distance along the plate
y = distance normal to the plate
u = x component of velocity
v = y component of velocity
ν = kinematic viscosity
α = thermal diffusivity

Equations (1.47) and (1.48) together with their boundary conditions, Eq. (1.50), constitute the well-known Blasius problem and can be transformed to

$$f''' + \tfrac{1}{2}ff'' = 0 \tag{1.51}$$

$$f(0)=f'(0)=0 \qquad f'(\infty)=1 \tag{1.52}$$

where

$$\begin{aligned} &f' = \frac{u}{U_\infty} \qquad v = \frac{1}{2}\left(\frac{\nu U_\infty}{x}\right)^{1/2}(\eta f' - f) \\ &\eta = \left(\frac{U_\infty}{\nu x}\right)^{1/2} y \end{aligned} \tag{1.53}$$

and primes denote differentiation with respect to η.

Considering the energy equation, we introduce two additional quantities θ and ϵ such that

$$\theta = \frac{T - T_\infty}{T_0 - T_\infty} \qquad \epsilon = \frac{U_\infty t}{x} \tag{1.54}$$

The energy equation and its associated conditions now become

$$\theta'' + \frac{1}{2}\,\mathrm{Pr}\,f\theta' = \mathrm{Pr}\,(1-\epsilon f')\,\frac{\partial\theta}{\partial\epsilon} \tag{1.55}$$

$$\begin{aligned} &\theta(0,\epsilon)=1 \qquad \theta(\infty,\epsilon)=0 \\ &\epsilon=0 \qquad \theta(\eta,\epsilon)=0 \end{aligned} \tag{1.56}$$

where $\mathrm{Pr} = \nu/\alpha$ is the Prandtl number. Here, the perturbation quantity ϵ represents dimensionless time and, as in Section 1.2.6, is a variable of the system. The assumption of small ϵ is appropriate during the early stages of thermal boundary-layer growth.

1.2.8 Cooling of a Lumped System with Variable Heat Transfer Coefficient

In the examples discussed so far, the perturbation quantity ϵ always carried a physical meaning. Now, we consider an example where ϵ is simply a small fraction and has no physical significance. Consider the cooling of a lumped system with temperature dependent heat transfer coefficient. Let the system have volume V, surface area A, density ρ, specific heat c, and initial temperature T_i. If the system is exposed at time $t = 0$ to an environment with heat transfer coefficient h and temperature T_a, the cooling of the system is described by the equation

$$\rho V c\,\frac{dT}{dt} = -hA(T - T_a) \tag{1.57}$$

$$t = 0 \qquad T = T_i \tag{1.58}$$

If the cooling is due to natural convection, then h takes the form

$$h = m(T - T_a)^{\epsilon} \tag{1.59}$$

where m is a constant and $\epsilon = \frac{1}{4}$ or $\frac{1}{3}$ depending upon whether the natural convection is laminar or turbulent.

Let us now introduce Eq. (1.59) and the following quantities

$$\theta = \frac{T - T_a}{T_i - T_a} \qquad \tau = \frac{Ah_i t}{\rho V c} \qquad h_i = m(T_i - T_a)^{\epsilon} \tag{1.60}$$

into Eqs. (1.57) and (1.58) to give

$$\frac{d\theta}{d\tau} + \theta^{1+\epsilon} = 0 \tag{1.61}$$

$$\tau = 0 \qquad \theta = 1 \tag{1.62}$$

1.3 PARAMETER PERTURBATION AND COORDINATE PERTURBATION

From the foregoing examples it is seen that the perturbation quantity can be a parameter or a coordinate. Accordingly, it is usual to classify the ensuing expansion as parameter perturbation or coordinate perturbation. Thus, expansions for examples shown in Sections 1.2.1-1.2.5 would be termed parameter expansions while those for examples given in Sections 1.2.6 and 1.2.7 would be termed coordinate expansions. The example in Section 1.2.8, however, does not fall in either of these categories.

1.4 CHOICE OF PERTURBATION QUANTITY

In most problems the perturbation quantity appears naturally in the equation and its choice is based on the physical understanding of the problem. However, it may sometimes become necessary to introduce one artificially to facilitate the expansion. Moreover, in the same problem, more than one choice may be available. The choice will then determine the simplicity or usefulness of the final solution. To illustrate this point, consider the problem of one-dimensional steady conduction in a slab with exponential heat generation and temperature dependent thermal conductivity. The governing equation is (Aziz and El-Ariny, 1977)

$$\frac{d}{dx}\left[k_0(1 + aT)\frac{dT}{dx}\right] + me^{T/T_s} = 0$$

where the second term is the heat generation contribution and $k_0(1 + aT)$ is the temperature-dependent heat conductivity.

By introducing the dimensionless variables

$$X = \frac{x}{L} \qquad \theta = \frac{T}{T_s}$$

and

$$\sigma = aT_s \qquad \beta = \frac{mL^2}{k_0 T_s}$$

the energy equation becomes

$$\frac{d}{dX}\left[(1 + \sigma\theta)\frac{d\theta}{dX}\right] + \beta e^{\theta} = 0 \tag{1.63}$$

Since both σ and β are small, the choice between σ and β as the perturbation quantity is open. However, the choice of β leads to a simpler perturbation solution (Aziz and El-Ariny, 1977). Similarly, the choice of $\epsilon = \nu(T_i - T_0)$ rather than νT_0 in the problem of transient heat conduction into a semi-infinite medium with variable heat capacity, renders the solution useful for larger ν and wider temperature range (Arunachalam and Seeniraj, 1977). Another example of judicious choice of perturbation quantity is provided by Aziz (1978) where the use of the thermal property parameter rather than the Stefan number leads to more useful solutions for freezing of a semi-infinite liquid with temperature-dependent thermal properties.

1.5 MULTIPLE PERTURBATION QUANTITIES

Most problems that are commonly encountered are characterized by the presence of one perturbation quantity. However, two or more quantities can appear in the same problem, each being appropriate as a choice for the perturbation quantity. Equation (1.63) is one such example. Here, we formulate another problem in detail to show how two perturbation quantities are identified.

1.5.1 Convecting–Radiating Fin with Variable Thermal Conductivity

Consider one-dimensional conduction in a straight fin of length L, cross-sectional area A, and perimeter P as shown in Fig. 1.7. The fin surface loses heat by combined convection and radiation. Let the environment temperature for convection be T_a and the effective sink temperature for radiation be T_s. Further, we impose the conditions of constant temperature T_b at the base and no heat loss from the tip.

Assuming the convective heat transfer coefficient h, and surface emissivity E_g to be constant, but allowing for variable thermal conductivity, the energy equation and the boundary conditions can be written as

$$\frac{d}{dx}\left(k\frac{dT}{dx}\right) - \frac{hP}{A}(T - T_a) - \frac{E_g\sigma}{A}(T^4 - T_s^4) = 0 \tag{1.64}$$

$$\begin{aligned} x &= 0 \quad \frac{dT}{dx} = 0 \\ x &= L \quad T = T_b \end{aligned} \tag{1.65}$$

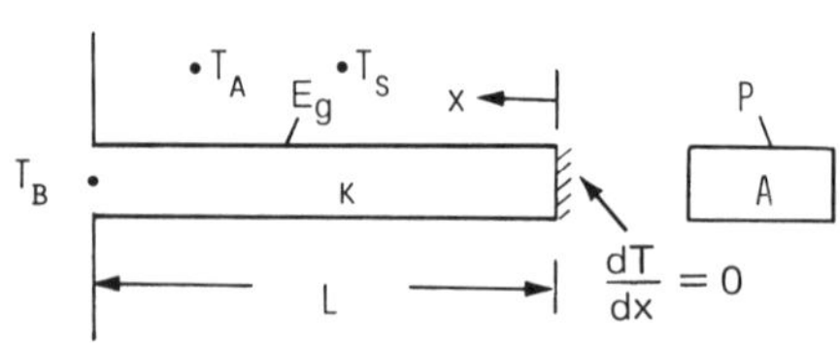

Figure 1.7 Convecting-radiating fin.

As in Section 1.2.1, assume k to be a linear function of T,

$$k = k_a[1 + \beta(T - T_a)] \tag{1.66}$$

Using Eq. (1.66) and introducing the following dimensionless quantities

$$\begin{aligned} &\theta = \frac{T}{T_b} \quad \theta_a = \frac{T_a}{T_b} \quad \theta_s = \frac{T_s}{T_b} \quad X = \frac{x}{L} \\ &N^2 = \frac{hPL^2}{k_a A} \quad \epsilon_1 = \beta T_b \quad \epsilon_2 = \frac{E_g \sigma T_b^3 PL^2}{k_a A} \end{aligned} \tag{1.67}$$

into Eqs. (1.64) and (1.65) gives

$$\frac{d}{dX}\left\{[1 + \epsilon_1(\theta - \theta_a)]\frac{d\theta}{dX}\right\} - N^2(\theta - \theta_a) - \epsilon_2(\theta^4 - \theta_s^4) = 0 \tag{1.68}$$

$$\begin{aligned} X = 0 \quad & \frac{d\theta}{dX} = 0 \\ X = 1 \quad & \theta = 1 \end{aligned} \tag{1.69}$$

Note that we have identified two perturbation quantities. The parameter ϵ_1 is the thermal conductivity parameter which appeared in slightly different form in Section 1.2.1. The parameter ϵ_2 is an indication of the role of radiation relative to conduction and may therefore be termed the radiation-conduction parameter.

1.6 GAUGE FUNCTIONS

Once a perturbation quantity ϵ has been identified, and stipulated to be small, the next step is to describe the dependence of the solution on ϵ. To lay the ground work for this development, we must know the behavior of a function $f(\epsilon)$ as ϵ approaches zero. Assuming that $f(\epsilon)$ has a quantitative limit, its limiting behavior can fall into one of three categories, namely,

$$\left.\begin{aligned} &\lim_{\epsilon \to 0} f(\epsilon) = 0 \\ &\lim_{\epsilon \to 0} f(\epsilon) = c \\ &\lim_{\epsilon \to 0} f(\epsilon) = \infty \end{aligned}\right\} \quad 0 < c < \infty \tag{1.70}$$

The foregoing classification is rather crude because a number of functions can fall under one category. To be more precise, we must know, at least qualitatively, the *rate* at which the limit is approached. For example, both $\sin \epsilon$ and $\operatorname{sech}(1/\epsilon)$ tend to zero as $\epsilon \to 0$, but their limiting behavior is different, as will be seen shortly. Therefore, for a more precise description, we compare the function $f(\epsilon)$ with a gauge function $g(\epsilon)$ whose limiting behavior is known to us. This comparison is facilitated by the use of order symbols O and o such that

$$f(\epsilon)=O[g(\epsilon)] \quad \text{as } \epsilon\to 0 \text{ if } \lim_{\epsilon\to 0}\frac{f(\epsilon)}{g(\epsilon)}<\infty \tag{1.71}$$

$$f(\epsilon)=o[g(\epsilon)] \quad \text{as } \epsilon\to 0 \text{ if } \lim_{\epsilon\to 0}\frac{f(\epsilon)}{g(\epsilon)}=0 \tag{1.72}$$

To illustrate the use of order symbol O we consider several examples.

First, let us determine the gauge function for $\sin\epsilon$ as $\epsilon\to 0$. From the Taylor series of $\sin\epsilon$, we have

$$\sin\epsilon=\epsilon-\frac{\epsilon^3}{3!}+\frac{\epsilon^5}{5!}-\frac{\epsilon^7}{7!}+\cdots$$

or

$$\frac{\sin\epsilon}{\epsilon}=1-\frac{\epsilon^2}{3!}+\frac{\epsilon^4}{5!}-\frac{\epsilon^6}{7!}+\cdots$$

As $\lim_{\epsilon\to 0}[(\sin\epsilon)/\epsilon]=1$, it is apparent that the limiting behavior of $\sin\epsilon$ is the same as ϵ which in this case is the appropriate gauge function. Hence,

$$\sin\epsilon=O(\epsilon) \quad \text{as } \epsilon\to 0 \tag{1.73}$$

Second, let us consider the limiting behavior of $1-\cos\epsilon$. Again, using the Taylor series for $\cos\epsilon$, we have

$$1-\cos\epsilon=\frac{\epsilon^2}{2!}-\frac{\epsilon^4}{4!}+\frac{\epsilon^6}{6!}-\frac{\epsilon^8}{8!}+\cdots$$

Since $\lim_{\epsilon\to 0}[(1-\cos\epsilon)/\epsilon^2]=\frac{1}{2}$, the appropriate gauge function is ϵ^2 and we can write

$$1-\cos\epsilon=O(\epsilon^2) \quad \text{as } \epsilon\to 0 \tag{1.74}$$

Third, consider the limiting behavior of $\cot\epsilon$ as $\epsilon\to 0$. The Taylor series gives

$$\cot\epsilon=\frac{1}{\epsilon}-\frac{\epsilon}{3}-\frac{\epsilon^3}{45}-\frac{2\epsilon^5}{945}-\cdots$$

Since $\lim_{\epsilon\to 0}[(\cot\epsilon)/(1/\epsilon)]=1$, the gauge function is $1/\epsilon$ or ϵ^{-1}. Thus,

$$\cot\epsilon=O(\epsilon^{-1}) \quad \text{as } \epsilon\to 0 \tag{1.75}$$

Fourth, let us determine the gauge function for $\epsilon^{3/2}/\sin\epsilon$. Since $\lim_{\epsilon\to 0}(\epsilon/\sin\epsilon)$ is also unity, it is obvious that $\epsilon^{1/2}$ is the correct gauge function. Thus,

$$\frac{\epsilon^{3/2}}{\sin \epsilon} = O(\epsilon^{1/2}) \quad \text{as } \epsilon \to 0 \tag{1.76}$$

These examples show that a useful set of gauge functions is $\epsilon^{\pm n}$ where n is an integer or a fraction. However, this set is not always adequate. For example, it fails to describe the behavior of $\sinh(1/\epsilon)$ as $\epsilon \to 0$. The function tends to infinity faster than any inverse power of ϵ because

$$\lim_{\epsilon \to 0} \frac{\sinh(1/\epsilon)}{\epsilon^n} = \lim_{s \to \infty} \frac{\sinh s}{s^n} = \lim_{s \to \infty} \frac{\sinh s}{n!} = \infty$$

The appropriate gauge function in this case is $e^{1/\epsilon}$ because

$$\lim_{\epsilon \to 0} \frac{\sinh(1/\epsilon)}{e^{1/\epsilon}} = \lim_{\epsilon \to 0} \frac{1 - e^{-2/\epsilon}}{2} = \frac{1}{2}$$

Hence,

$$\sinh \frac{1}{\epsilon} = O(e^{1/\epsilon}) \quad \text{as } \epsilon \to 0 \tag{1.77}$$

With a similar argument, it can be shown that

$$\operatorname{sech} \frac{1}{\epsilon} = O(e^{-1/\epsilon}) \quad \text{as } \epsilon \to 0 \tag{1.78}$$

The conclusion is that for functions which tend to infinity faster than any inverse power of ϵ, the gauge function is $e^{1/\epsilon}$. Likewise, for functions which tend to zero faster than any direct power of ϵ, the gauge function is $e^{-1/\epsilon}$.

To explore further possibilities, let us try to describe the behavior of $\operatorname{sech}^{-1} \epsilon$ as $\epsilon \to 0$. The function tends to infinity, as $\epsilon \to 0$, more slowly than any inverse power of ϵ because $\lim_{\epsilon \to 0} [(\operatorname{sech}^{-1} \epsilon)/\epsilon^{-n}] = 0$. In this case, the appropriate gauge function is $\ln(1/\epsilon)$ because

$$\lim_{\epsilon \to 0} \frac{\operatorname{sech}^{-1} \epsilon}{\ln(1/\epsilon)} = \lim_{\epsilon \to 0} \frac{\ln(1/\epsilon) + \ln(1 + \sqrt{1 - \epsilon^2})}{\ln(1/\epsilon)} = 1$$

Hence,

$$\operatorname{sech}^{-1} \epsilon = O\left(\ln \frac{1}{\epsilon}\right) \quad \text{as } \epsilon \to 0 \tag{1.79}$$

It may likewise be shown that $[\ln(1/\epsilon)]^{-1}$ is the correct gauge function for functions that tend to zero more slowly than any direct power of ϵ.

In summary, it can be concluded that a complete set of gauge functions involves not only powers of ϵ but also its exponential, logarithm, log log, etc. However, the most useful is the simple set $\epsilon^{\pm n}$, which is used almost exclusively in this book.

1.7 ASYMPTOTIC EXPANSIONS

Having introduced the concept of the gauge function, we now wish to express $f(\epsilon)$ as a function of ϵ. First, let us define a sequence of gauge functions $g_n(\epsilon)$, called an asymptotic sequence, such that

$$g_n(\epsilon) = o[g_{n-1}(\epsilon)] \tag{1.80}$$

Next, we write an asymptotic series expansion for $f(\epsilon)$ using the sequence $g_n(\epsilon)$:

$$f(\epsilon) = \sum_{n=0}^{\infty} a_n g_n(\epsilon) \tag{1.81}$$

A special form of Eq. (1.81) is

$$f(\epsilon) = \sum_{n=0}^{\infty} a_n \epsilon^n \tag{1.82}$$

where the asymptotic sequence is ϵ^n. This is the most commonly used form of asymptotic expansion.

Where multiple perturbation quantities ϵ_1, ϵ_2, . . . , etc. appear, Eq. (1.81) takes the form of a multiple series. For example, the double series version of Eq. (1.82) involving ϵ_1 and ϵ_2 is

$$f(\epsilon_1, \epsilon_2) = \sum_{m=0}^{\infty} \sum_{n=0}^{\infty} a_{mn} \epsilon_1^m \epsilon_2^n \tag{1.83}$$

1.8 REGULAR PERTURBATION AND SINGULAR PERTURBATION

Having identified a perturbation quantity and chosen the form of the asymptotic expansion, we now come to the main task which is to determine the coefficients a_n in Eq. (1.82). The procedure is to substitute Eq. (1.82) into the governing equations of the problem. For example, if we wish to solve Eq. (1.5) subject to the boundary conditions in Eq. (1.6), we assume

$$\theta = \sum_{n=0}^{\infty} a_n \epsilon^n$$

and substitute it in Eqs. (1.5) and (1.6). To carry out this substitution, we would, in general, need to perform operations such as addition, subtraction, multiplication, exponentiation, differentiation, and integration on the asymptotic expansion in Eq. (1.82). The justification of these operations is discussed by Erdélyi (1956). Rather than delve into the subject, we perform these operations in the normal manner assuming they are justified.

After substitution, the next step involves collection of terms having like powers of ϵ. Equating the coefficient of each power of ϵ to zero, one is led to a sequence of problems (instead of one original problem) which can be solved *in succession* to obtain the coefficients a_n. This process of determining a_n will be discussed in depth in the following chapter. With the evaluation of a_n, the asymptotic expansion in Eq. (1.82) is fully determined. Next, we examine its validity in the entire domain of the independent variable. If valid throughout the domain, the expansion is classified as *regular perturbation expansion.* On the other hand, if it breaks down (i.e., becomes singular) in a certain region, the expansion is classified as *singular perturbation expansion.* The next chapter deals with regular perturbation expansions while the subsequent three chapters are devoted to singular perturbation expansions.

CHAPTER

TWO

REGULAR PERTURBATION EXPANSIONS

2.1 INTRODUCTION

This chapter aims to provide a thorough exposition of regular perturbation technique. The problems chosen for discussion cover a wide spectrum of heat transfer theory. We begin with the solution of algebraic and transcendental equations, then move to ordinary differential equations, and finally deal with partial differential equations. With the first few problems, the analysis will be presented in full detail so that the reader can grasp the technique without difficulty. However, as the ideas are gradually consolidated, the routine details will be omitted, and attention will be focused on the core of the analysis.

2.2 AN ALGEBRAIC EQUATION

Consider the simple quadratic equation

$$x^2 - (3 + \epsilon)x + 2 = 0 \tag{2.1}$$

where ϵ is small. We wish to determine the two roots using perturbation expansion. Let us assume a three-term expansion of the form

$$x = x_0 + \epsilon x_1 + \epsilon^2 x_2 + O(\epsilon^3) \tag{2.2}$$

and substitute Eq. (2.2) into Eq. (2.1). This gives

$$(x_0 + \epsilon x_1 + \epsilon^2 x_2 + \cdots)^2 - (3+\epsilon)(x_0 + \epsilon x_1 + \epsilon^2 x_2 + \cdots) + 2 = 0 \tag{2.3}$$

Expanding Eq. (2.3), we write

$$\begin{aligned} x_0^2 + \epsilon 2x_0x_1 + \epsilon^2(2x_0x_2 + x_1^2) + \cdots - 3x_0 - \epsilon(x_0 + 3x_1) \\ - \epsilon^2(x_1 + 3x_2) - \cdots + 2 = 0 \end{aligned} \tag{2.4}$$

We now collect coefficients of like powers of ϵ and write Eq. (2.4) as

$$(x_0^2 - 3x_0 + 2) + \epsilon(2x_0x_1 - x_0 - 3x_1) + \epsilon^2(2x_0x_2 + x_1^2 - x_1 - 3x_2) = 0 \tag{2.5}$$

Since Eq. (2.5) is an identity in ϵ, the coefficient of each power of ϵ must vanish. By equating the coefficient of ϵ^0 with zero, we get

$$x_0^2 - 3x_0 + 2 = 0 \tag{2.6}$$

Equation (2.6) is termed the *zero-order problem* and its solution

$$x_0 = 1 \text{ or } 2 \tag{2.7}$$

is accordingly called the *zero-order solution*.

Similarly, by equating the coefficient of ϵ with zero, we obtain the *first-order problem*

$$2x_0x_1 - x_0 - 3x_1 = 0 \tag{2.8}$$

whose solution, using Eq. (2.7), is

$$x_1 = -1 \text{ or } 2 \tag{2.9}$$

Equation (2.9) is the *first-order solution*.

Finally, we equate the coefficient of ϵ^2 with zero to obtain the *second-order problem*

$$2x_0x_2 + x_1^2 - 3x_2 - x_1 = 0 \tag{2.10}$$

whose solution, using Eqs. (2.7) and (2.9), is

$$x_2 = 2 \text{ or } -2 \tag{2.11}$$

Equation (2.11) is the *second-order solution*.

Thus the expansion shown in Eq. (2.2) gives the two roots as

$$x = 1 - \epsilon + 2\epsilon^2 \tag{2.12}$$

and

$$x = 2 + 2\epsilon - 2\epsilon^2 \tag{2.13}$$

It should be noted that the perturbation equations for x_0, x_1, x_2, etc. are solved *in succession*. It is readily seen that the exact solution of Eq. (2.1) is

$$x = \tfrac{1}{2}(3 + \epsilon \pm \sqrt{1 + 6\epsilon + \epsilon^2}) \tag{2.14}$$

Expanding $\sqrt{1 + 6\epsilon + \epsilon^2}$ binomially gives $1 + 3\epsilon - 4\epsilon^2 + O(\epsilon^3)$. Thus with plus sign, Eq. (2.14) reproduces Eq. (2.13); with minus sign, it reproduces Eq. (2.12). Therefore, in this example, the perturbation expansion reproduces the exact solution.

2.3 A TRANSCENDENTAL EQUATION

Consider the problem of transient heat conduction in a slab in which heat is generated at a rate which increases exponentially with temperature (Carslaw and Jaeger, 1959). To solve for the steady state temperature distribution, it is necessary to compute the root of the transcendental equation

$$z = \tfrac{1}{2}\beta \cosh z \tag{2.15}$$

where β is the heat generation parameter. Since the steady state exists if $\beta < 0.88$, we may define $\epsilon = \frac{1}{2}\beta$ and assume it to be small. Following Aziz (1977) we assume

$$z = z_0 + \epsilon z_1 + \epsilon^2 z_2 + \epsilon^3 z_3 + O(\epsilon^4) \tag{2.16}$$

and calculate the first four coefficients z_0, z_1, z_2, and z_3.

Substituting Eq. (2.16) into Eq. (2.15), we have

$$z_0 + \epsilon z_1 + \epsilon^2 z_2 + \epsilon^3 z_3 + \cdots = \epsilon \cosh (z_0 + \epsilon z_1 + \epsilon^2 z_2 + \epsilon^3 z_3 + \cdots) \tag{2.17}$$

To obtain an explicit expression on the right-hand side of Eq. (2.17) we use the expansion for $\cosh z$

$$\cosh z = 1 + \frac{z^2}{2!} + \frac{z^4}{4!} + \cdots \tag{2.18}$$

Using Eq. (2.18), Eq. (2.17) becomes

$$z_0 + \epsilon z_1 + \epsilon^2 z_2 + \epsilon^3 z_3 + \cdots = \epsilon \left[1 + \frac{(z_0 + \epsilon z_1 + \epsilon^2 z_2 + \epsilon^3 z_3 + \cdots)^2}{2!} + \frac{(z_0 + \epsilon z_1 + \epsilon^2 z_2 + \epsilon^3 z_3 + \cdots)^4}{4!} + \cdots \right] \tag{2.19}$$

Using binomial expansion and retaining terms up to $O(\epsilon^2)$, we have

$$(z_0 + \epsilon z_1 + \epsilon^2 z_2 + \epsilon^3 z_3 + \cdots)^2 = z_0^2 + \epsilon 2 z_0 z_1 + \epsilon^2 (2 z_0 z_2 + z_1^2) + \cdots \tag{2.20}$$

$$(z_0 + \epsilon z_1 + \epsilon^2 z_2 + \epsilon^3 z_3 + \cdots)^4 = z_0^4 + \epsilon 4 z_0^3 z_1 + \epsilon^2 (4 z_0^3 z_2 + 6 z_0^2 z_1^2) + \cdots \tag{2.21}$$

Utilizing Eqs. (2.20) and (2.21) in Eq. (2.19), and collecting coefficients of like powers of ϵ, we get

$$z_0 + \epsilon z_1 + \epsilon^2 z_2 + \epsilon^3 z_3 + \cdots = \epsilon(1 + z_0^2 + z_0^4) + \epsilon^2 (z_0 z_1 + \tfrac{1}{6} z_0^3 z_1) + \epsilon^3 (z_0 z_2 + \tfrac{1}{2} z_1^2 + \tfrac{1}{6} z_0^3 z_2 + \tfrac{1}{4} z_0^2 z_1^2) + \cdots \tag{2.22}$$

Equating the coefficients of each power of ϵ on both sides of Eq. (2.22) gives

$$\epsilon^0: \quad z_0 = 0 \tag{2.23}$$

$$\epsilon^1: \quad z_1 = 1 + z_0^2 + z_0^4 \tag{2.24}$$

$$\epsilon^2: \quad z_2 = z_0 z_1 + \tfrac{1}{6} z_0^3 z_1 \tag{2.25}$$

$$\epsilon^3: \quad z_3 = z_0 z_2 + \tfrac{1}{2} z_1 + \tfrac{1}{6} z_0^3 z_2 + \tfrac{1}{4} z_0^2 z_1^2 \tag{2.26}$$

The solutions of Eqs. (2.24)-(2.26) are

$$z_1 = 1 \qquad z_2 = 0 \qquad z_3 = \tfrac{1}{2} \tag{2.27}$$

Thus the expansion in Eq. (2.16) becomes

$$z = \epsilon + \tfrac{1}{2}\epsilon^3 + \cdots \tag{2.28}$$

To assess the accuracy of Eq. (2.28), we compare it in Table 2.1 with the iterative solution of Eq. (2.15).

Examining Table 2.1 it is seen that for small β, the error between the perturbation and the iterative solution is small. As β increases the error in-

Table 2.1 Solutions of $z = \frac{1}{2}\beta \cosh z$

β	z (perturbation)	z (iteration)
0.1	0.2292	0.2295
0.2	0.3320	0.3340
0.3	0.4163	0.4223
0.4	0.4919	0.5050
0.5	0.5625	0.5890
0.6	0.6298	0.6788
0.7	0.6951	0.7818
0.8	0.7589	0.9210
0.88	0.8092	1.1220

creases. For example, at $\beta = 0.5$, the error is about 4 percent, but at $\beta = 0.88$, it is about 28 percent. Thus the usefulness of Eq. (2.28) is restricted to small values of ϵ or β. However, the margin of error can be reduced by calculating additional terms of Eq. (2.28) as shown in Problem 2.2.

It is interesting to note that Eq. (2.15) can be expanded using Eq. (2.18) to give ϵ as a power series in z. The latter, when reversed, leads exactly to the perturbation series of Eq. (2.28).

Equation (2.15) has two roots but the perturbation solution gives only the smaller root. However, this does not bother us because it is the smaller root that gives a steady solution and is therefore of interest to us.

2.4 COOLING OF A LUMPED SYSTEM WITH VARIABLE SPECIFIC HEAT

Consider the cooling of a lumped system. Let the system have volume V, surface area A, density ρ, specific heat c and initial temperature T_i. At time $t = 0$, the system is exposed to a convective environment at temperature T_a with convective heat transfer coefficient h. Assume that the specific heat c is a linear function temperature of the form

$$c = c_a[1 + \beta(T - T_a)] \tag{2.29}$$

where c_a is the specific heat at temperature T_a and β is a constant. The cooling equation and the initial conditions are

$$\rho V c \frac{dT}{dt} + hA(T - T_a) = 0 \tag{2.30}$$

$$t = 0 \qquad T = T_i \tag{2.31}$$

Introducing Eq. (2.29) and using the dimensionless parameters

$$\theta = \frac{T - T_a}{T_i - T_a} \qquad \tau = \frac{t}{\rho V c_a/(hA)} \qquad \epsilon = \beta(T_i - T_a)$$

transforms Eq. (2.30) to

$$(1 + \epsilon\theta)\frac{d\theta}{d\tau} + \theta = 0 \tag{2.32}$$

$$\tau = 0 \qquad \theta = 0 \tag{2.33}$$

For $\epsilon \ll 1$, let us assume a regular perturbation expansion and calculate the first three terms (Aziz and Hamad, 1977). Thus we assume

$$\theta = \theta_0 + \epsilon\theta_1 + \epsilon^2\theta_2 + \cdots \tag{2.34}$$

Substituting Eq. (2.34) into Eqs. (2.32) and (2.33) and carrying out the elementary operations, we have

$$[1+\epsilon(\theta_0+\epsilon\theta_1+\epsilon^2\theta_2)]\left(\frac{d\theta_0}{d\tau}+\epsilon\frac{d\theta_1}{d\tau}+\epsilon^2\frac{d\theta_2}{d\tau}\right)$$
$$+\theta_0+\epsilon\theta_1+\epsilon^2\theta_2=0 \tag{2.35}$$

$$\tau=0 \qquad \theta_0+\epsilon\theta_1+\epsilon^2\theta_2=1 \tag{2.36}$$

Expanding Eq. (2.35) and collecting terms with like powers of ϵ gives

$$\frac{d\theta_0}{d\tau}+\theta_0+\epsilon\left(\frac{d\theta_1}{d\tau}+\theta_1+\theta_0\frac{d\theta_0}{d\tau}\right)$$
$$+\epsilon^2\left(\frac{d\theta_2}{d\tau}+\theta_2+\theta_0\frac{d\theta_1}{d\tau}+\theta_1\frac{d\theta_0}{d\tau}\right)=0 \tag{2.37}$$

Equating coefficients of each power of ϵ on both sides of Eqs. (2.37) and (2.36) gives

$$\epsilon^0: \qquad \frac{d\theta_0}{d\tau}+\theta_0=0 \tag{2.38}$$

$$\tau=0 \qquad \theta_0=1 \tag{2.39}$$

$$\epsilon^1: \qquad \frac{d\theta_1}{d\tau}+\theta_1+\theta_0\frac{d\theta_0}{d\tau}=0 \tag{2.40}$$

$$\tau=0 \qquad \theta_1=0 \tag{2.41}$$

$$\epsilon^2: \qquad \frac{d\theta_2}{d\tau}+\theta_2+\theta_0\frac{d\theta_1}{d\tau}+\theta_1\frac{d\theta_0}{d\tau}=0 \tag{2.42}$$

$$\tau=0 \qquad \theta_2=0 \tag{2.43}$$

The solution for θ_0 is

$$\theta_0=e^{-\tau} \tag{2.44}$$

which is substituted in Eq. (2.40) to give

$$\frac{d\theta_1}{d\tau}+\theta_1-e^{-2\tau}=0 \tag{2.45}$$

The solution of Eq. (2.45) subject to Eq. (2.41) is

$$\theta_1=e^{-\tau}-e^{-2\tau} \tag{2.46}$$

Finally, using Eqs. (2.44) and (2.46), Eq. (2.42) becomes

$$\frac{d\theta_2}{d\tau}+\theta_2-2e^{-2\tau}+3e^{-3\tau}=0 \tag{2.47}$$

The solution of Eq. (2.47) subject to Eq. (2.43) is

$$\theta_2 = e^{-\tau} - 2e^{-2\tau} + \tfrac{3}{2}e^{-3\tau} \tag{2.48}$$

The three-term expansion in Eq. (2.34) now becomes

$$\theta = e^{-\tau} + \epsilon(e^{-\tau} - e^{-2\tau}) + \epsilon^2(e^{-\tau} - 2e^{-2\tau} + \tfrac{3}{2}e^{-3\tau}) \tag{2.49}$$

By separating the variables in Eq. (2.32) and carrying out the integration, the exact solution can be obtained as

$$\ln\theta + \epsilon(\theta - 1) = -\tau \tag{2.50}$$

In Fig. 2.1 we compare the perturbation solution in Eq. (2.49) with the exact solution in Eq. (2.50) for $\epsilon = 0$, $+0.2$, and -0.2. The agreement is excellent.

2.5 PLANE COUETTE FLOW WITH VARIABLE VISCOSITY

This problem was formulated in Section 1.2.2. The governing equations are Eqs. (1.13) and (1.14) with boundary conditions shown in Eq. (1.15). Let us derive a three-term perturbation solution by assuming

$$U = U_0 + \epsilon U_1 + \epsilon^2 U_2 \tag{2.51}$$

$$\theta = \theta_0 + \epsilon\theta_1 + \epsilon^2\theta_2 \tag{2.52}$$

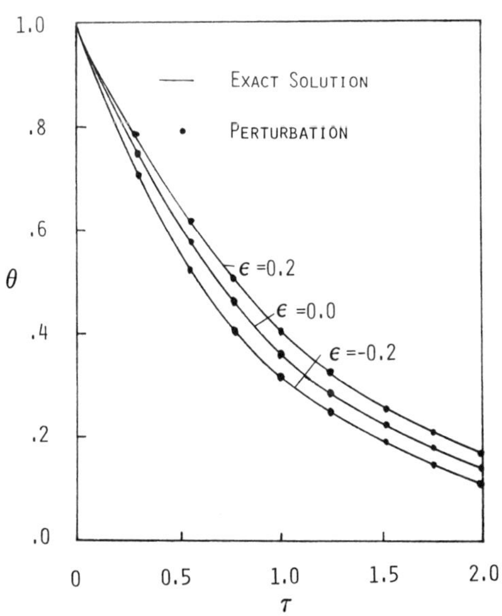

Figure 2.1 Cooling of a lumped system.

Substituting Eqs. (2.51) and (2.52) into Eqs. (1.13) and (1.14) we encounter the term $\exp[-\beta(\theta_0 + \epsilon\theta_1 + \epsilon^2\theta_2)]$ which needs to be expanded before coefficients of like powers of ϵ can be collected together. To this effect, we write

$$\exp[-\beta(\theta_0 + \epsilon\theta_1 + \epsilon^2\theta_2)] = \exp(-\beta\theta_0)\exp(-\epsilon\beta\theta_1)\exp(-\epsilon^2\beta\theta_2)\cdots$$

and expand the individual term, retaining terms up to ϵ^2. Thus

$$\begin{aligned}\exp[-\beta(\theta_0 + \epsilon\theta_1 + \epsilon^2\theta_2)] &= e^{-\beta\theta_0}\left[1 - \epsilon(\beta\theta_1) + \epsilon^2\frac{(\beta\theta_1)^2}{2}\right](1 - \epsilon^2\beta\theta_2)\\ &= e^{-\beta\theta_0}\left\{1 - \epsilon(\beta\theta_1) + \epsilon^2\left[\frac{(\beta\theta_1)^2}{2} - \beta\theta_2\right]\right\}\\ &= e^{-\beta\theta_0} - \epsilon e^{-\beta\theta_0}\beta\theta_1 + \epsilon^2 e^{-\beta\theta_0}\left[\frac{(\beta\theta_1)^2}{2} - \beta\theta_2\right]\end{aligned} \tag{2.53}$$

Similarly the term $(dU/dY)^2$ must be expanded as follows

$$\left(\frac{dU}{dY}\right)^2 = \left(\frac{dU_0}{dY} + \epsilon\frac{dU_1}{dY} + \epsilon^2\frac{dU_2}{dY}\right)^2 = \left(\frac{dU_0}{dY}\right)^2 + \epsilon\left(2\frac{dU_0}{dY}\frac{dU_1}{dY}\right) + \epsilon^2\left[2\frac{dU_0}{dY}\frac{dU_1}{dY} + \left(\frac{dU_1}{dY}\right)^2\right] \tag{2.54}$$

Making use of Eqs. (2.53) and (2.54), the sequence of perturbation equations follows as

ϵ^0:

$$\frac{d}{dY}\left(e^{-\beta\theta_0}\frac{dU_0}{dY}\right) = 0 \tag{2.55}$$

$$\frac{d^2\theta_0}{dY^2} = 0 \tag{2.56}$$

$$Y = 0 \quad U_0 = 0 \quad \theta_0 = 0 \tag{2.57}$$

$$Y = 1 \quad U_0 = 1 \quad \theta_0 = 0 \tag{2.58}$$

ϵ^1:

$$\frac{d}{dY}\left[e^{-\beta\theta_0}\left(\frac{dU_1}{dY} - \beta\theta_1\frac{dU_0}{dY}\right)\right] = 0 \tag{2.59}$$

$$\frac{d^2\theta_1}{dY^2} + e^{-\beta\theta_0}\left(\frac{dU_0}{dY}\right)^2 = 0 \tag{2.60}$$

$$Y = 0 \quad U_1 = 0 \quad \theta_1 = 0 \tag{2.61}$$

$$Y = 1 \quad U_1 = 0 \quad \theta_1 = 0 \tag{2.62}$$

$$\epsilon^2: \quad \frac{d}{dY}\left(e^{-\beta\theta_0}\left\{\frac{dU_2}{dY} - \beta\theta_1 \frac{dU_1}{dY} + \left[\frac{(\beta\theta_1)^2}{2} - \beta\theta_2\right]\frac{dU_0}{dY}\right\}\right) = 0 \tag{2.63}$$

$$\frac{d^2\theta_2}{dY^2} + e^{-\beta\theta_0}\left[2\frac{dU_0}{dY}\frac{dU_1}{dY} - \beta\theta_1\left(\frac{dU_0}{dY}\right)^2\right] = 0 \tag{2.64}$$

$$Y = 0 \quad U_2 = 0 \quad \theta_2 = 0 \tag{2.65}$$

$$Y = 1 \quad U_2 = 0 \quad \theta_2 = 0 \tag{2.66}$$

It is evident that the solution sequence should be θ_0, U_0, θ_1, U_1, θ_2, and U_2 in that order. The integration involved in solving the sequence of equations is elementary. The final results are

$$\theta_0 = 0 \tag{2.67}$$

$$U_0 = Y \tag{2.68}$$

$$\theta_1 = \tfrac{1}{2}Y(1 - Y) \tag{2.69}$$

$$U_1 = -\tfrac{1}{12}\beta(Y - 3Y^2 + 2Y^3) \tag{2.70}$$

$$\theta_2 = -\tfrac{1}{24}\beta(Y - 2Y^2 + 2Y^3 - Y^4) \tag{2.71}$$

$$U_2 = \tfrac{1}{120}\beta^2(Y - 5Y^2 + 10Y^3 - 10Y^4 + 4Y^5) \tag{2.72}$$

The foregoing perturbation solution may be compared with the exact solution (Turian and Bird, 1963) which is given by

$$e^{\beta\theta} = \left(1 + \frac{\epsilon\beta}{8}\right)\operatorname{sech}^2\left[(2Y - 1)\sinh^{-1}\left(\frac{\epsilon\beta}{8}\right)^{1/2}\right] \tag{2.73}$$

$$U = \frac{1}{2}\left\{\left(1 + \frac{8}{\epsilon\beta}\right)^{1/2}\tanh\left[(2Y - 1)\sinh^{-1}\left(\frac{\epsilon\beta}{8}\right)^{1/2}\right] + 1\right\} \tag{2.74}$$

The comparison is made in Fig. 2.2 for $\beta = 1$ and a range of values of ϵ. Once again the predictions from the perturbation solution are quite good.

2.6 LAMINAR MIXED CONVECTION IN A VERTICAL PIPE

Equations (1.20)-(1.22) derived in Section 1.2.3 will be solved here. Assume a regular perturbation expansion as

$$U = U_0 + \epsilon U_1 + \epsilon^2 U_2 + O(\epsilon^3) \tag{2.75}$$

$$\theta = \theta_0 + \epsilon\theta_1 + \epsilon^2\theta_2 + O(\epsilon^3) \tag{2.76}$$

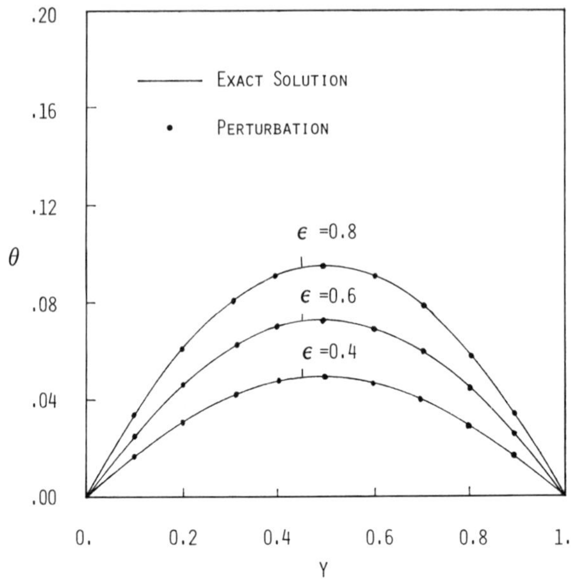

Figure 2.2 Temperature distribution in plane couette flow, $\beta = 1$.

Substituting Eqs. (2.75) and (2.76) into Eqs. (1.20)-(1.22) and following the usual procedure, we get the following equations for U_0, U_1, U_2, θ_0, θ_1, and θ_2

$$\epsilon^0: \quad \frac{d^2U_0}{dR^2} + \frac{1}{R}\frac{dU_0}{dR} = -P \tag{2.77}$$

$$\frac{d^2\theta_0}{dR^2} + \frac{1}{R}\frac{d\theta_0}{dR} = -U_0 \tag{2.78}$$

$$\epsilon^1: \quad \frac{d^2U_1}{dR^2} + \frac{1}{R}\frac{dU_1}{dR} = \theta_0 \tag{2.79}$$

$$\frac{d^2\theta_1}{dR^2} + \frac{1}{R}\frac{d\theta_1}{dR} = -U_1 \tag{2.80}$$

$$\epsilon^2: \quad \frac{d^2U_2}{dR^2} + \frac{1}{R}\frac{dU_2}{dR} = \theta_1 \tag{2.81}$$

$$\frac{d^2\theta_2}{dR^2} + \frac{1}{R}\frac{d\theta_2}{dR} = -U_2 \tag{2.82}$$

subject to the boundary conditions

$$R = 0 \quad \frac{dU_0}{dR} = \frac{dU_1}{dR} = \frac{dU_2}{dR} = \frac{d\theta_0}{dR} = \frac{d\theta_1}{dR} = \frac{d\theta_2}{dR} = 0 \tag{2.83}$$

$$R = 1 \qquad U_0 = U_1 = U_2 = \theta_0 = \theta_1 = \theta_2 = 0 \tag{2.84}$$

Since the integration of Eqs. (2.77)-(2.82) is elementary, we give the final results, which are

$$U_0 = \frac{P}{4}(1 - R^2) \tag{2.85}$$

$$\theta_0 = \frac{P}{64}(R^4 - 4R^2 + 3) \tag{2.86}$$

$$U_1 = \frac{P}{2{,}304}(R^6 - 9R^4 + 27R^2 - 19) \tag{2.87}$$

$$\theta_1 = \frac{P}{147{,}456}(R^8 + 16R^6 - 108R^4 + 304R^2 - 211) \tag{2.88}$$

$$U_2 = -\frac{P}{14{,}745{,}600}(R^{10} - 25R^8 + 300R^6 - 1{,}900R^4 + 5{,}275R^2 - 3{,}651) \tag{2.89}$$

$$\theta_2 = \frac{P}{2{,}123{,}366{,}400}(R^{12} - 36R^{10} + 675R^8 - 7{,}600R^6 + 47{,}475R^4 - 131{,}436R^2 + 90{,}921) \tag{2.90}$$

To assess the accuracy of the foregoing solution, we can compare it with the exact solution of Eqs. (1.20)-(1.22) which is

$$U = P\epsilon^{-1/2}\frac{\operatorname{bei}\epsilon^{1/4}\operatorname{ber}\epsilon^{1/4}R - \operatorname{ber}\epsilon^{1/4}\operatorname{bei}\epsilon^{1/4}R}{\operatorname{ber}^2\epsilon^{1/4} + \operatorname{bei}^2\epsilon^{1/4}} \tag{2.91}$$

$$\theta = P\epsilon^{-1}\left(1 - \frac{\operatorname{ber}\epsilon^{1/4}\operatorname{ber}\epsilon^{1/4}R + \operatorname{bei}\epsilon^{1/4}\operatorname{bei}\epsilon^{1/4}R}{\operatorname{ber}^2\epsilon^{1/4} + \operatorname{bei}^2\epsilon^{1/4}}\right) \tag{2.92}$$

where ber and bei are Kelvin functions. In Fig. 2.3 the velocity profile is plotted for a range of values of $\epsilon^{1/4}$. The corresponding plots for the temperature profile appear in Fig. 2.4.

2.7 FREEZING OF A SATURATED LIQUID IN SEMI-INFINITE REGION

The mathematical model for this problem was derived in Section 1.2.4 and is represented by Eqs. (1.27)-(1.29). We derive a three-term perturbation solution by assuming

$$\theta = \theta_0 + \epsilon\theta_1 + \epsilon^2\theta_2 \tag{2.93}$$

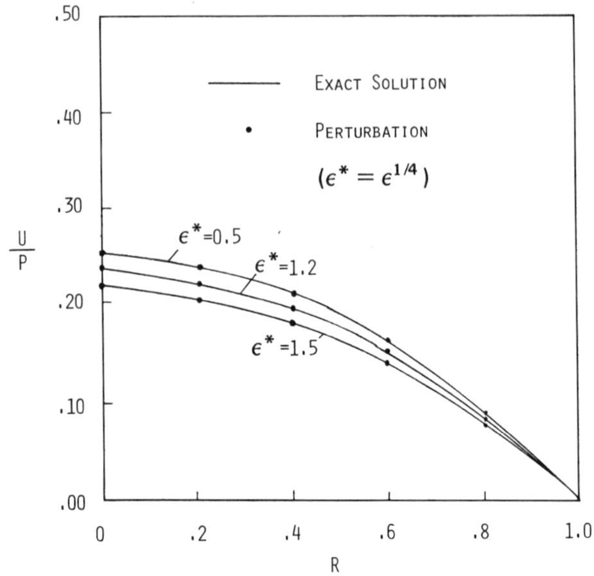

Figure 2.3 Velocity distribution in mixed convection in a vertical pipe.

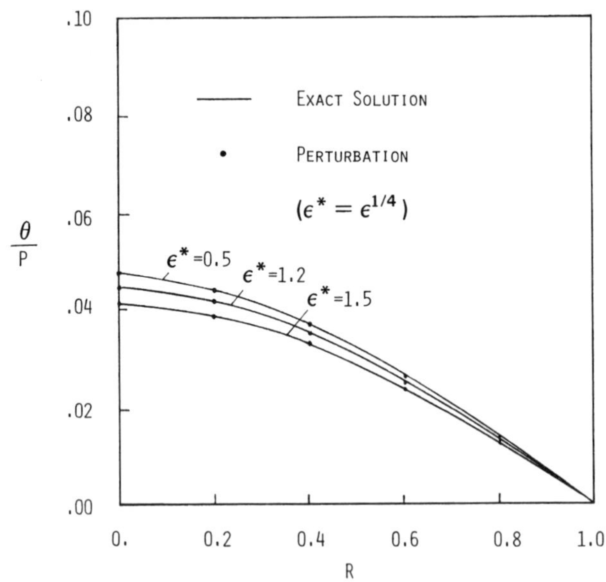

Figure 2.4 Temperature distribution in mixed convection in a vertical pipe.

Proceeding in the usual manner, the governing equations for θ_0, θ_1, and θ_2 are

$$\epsilon^0: \quad \frac{\partial^2 \theta_0}{\partial X^2} = 0 \tag{2.94}$$

$$\theta_0(X=0, X_f) = 1 \qquad \theta_0(X=X_f, X_f) = 0 \tag{2.95}$$

$$\epsilon^1: \quad \frac{\partial^2 \theta_1}{\partial X^2} = -\frac{\partial \theta_0}{\partial X_f}\frac{\partial \theta_0}{\partial X}\bigg|_{X=X_f} \tag{2.96}$$

$$\theta_1(X=0, X_f) = 0 \qquad \theta_1(X=X_f, X_f) = 0 \tag{2.97}$$

$$\epsilon^2: \quad \frac{\partial^2 \theta_2}{\partial X^2} = -\left(\frac{\partial \theta_0}{\partial X_f}\frac{\partial \theta_1}{\partial X}\bigg|_{X=X_f} + \frac{\partial \theta_1}{\partial X_f}\frac{\partial \theta_0}{\partial X}\bigg|_{X=X_f}\right) \tag{2.98}$$

$$\theta_2(X=0, X_f) = 0 \qquad \theta_2(X=X_f, X_f) = 0 \tag{2.99}$$

Integrating Eq. (2.94) twice, we have

$$\theta_0 = f_1 X + f_2$$

where f_1 and f_2 are functions of X_f. Applying the boundary conditions in Eq. (2.95) gives $f_1 = -1/X_f$ and $f_2 = 1$. Thus

$$\theta_0 = 1 - \frac{X}{X_f} \tag{2.100}$$

Using Eq. (2.100) on the right-hand side of Eq. (2.96) gives

$$\frac{\partial^2 \theta_1}{\partial X^2} = \frac{X}{X_f^3} \tag{2.101}$$

Integrating twice, we have

$$\theta_1 = \frac{1}{6}\left(\frac{X}{X_f}\right)^3 + f_3 X + f_4$$

where f_3 and f_4 are functions of X_f. Applying the boundary conditions in Eq. (2.97) gives $f_3 = -1/6X_f$ and $f_4 = 0$. Thus

$$\theta_1 = -\frac{1}{6}\frac{X}{X_f}\left[1 - \left(\frac{X}{X_f}\right)^2\right] \tag{2.102}$$

Now we use Eqs. (2.100) and (2.102) in Eq. (2.98) to obtain

$$\frac{\partial^2 \theta_2}{\partial X^2} = -\left(\frac{1}{6}\frac{X}{X_f^3} + \frac{1}{2}\frac{X^3}{X_f^5}\right)$$

Integrating twice, we have

$$\theta_2 = -\left[\frac{1}{36}\left(\frac{X}{X_f}\right)^3 + \frac{1}{40}\left(\frac{X}{X_f}\right)^5\right] + f_5 X + f_6$$

where f_5 and f_6 are functions of X_f. Applying the boundary conditions in Eq. (2.99) gives $f_5 = 19/(360X_f)$ and $f_6 = 0$. Thus

$$\theta_2 = \frac{1}{360}\frac{X}{X_f}\left[19 - 10\left(\frac{X}{X_f}\right)^2 - 9\left(\frac{X}{X_f}\right)^4\right] \tag{2.103}$$

The final three-term expansion now takes the form

$$\theta = 1 - \frac{X}{X_f} - \frac{1}{6}\epsilon\frac{X}{X_f}\left[1 - \left(\frac{X}{X_f}\right)^2\right] + \frac{1}{360}\epsilon^2\frac{X}{X_f}\left[19 - 10\left(\frac{X}{X_f}\right)^2 - 9\left(\frac{X}{X_f}\right)^4\right] + O(\epsilon^3) \tag{2.104}$$

We can now derive an expression for the progress of the freezing front with time. Using Eq. (2.104) in Eq. (1.29), it is easy to find that

$$\frac{dX_f}{d\tau} = \frac{1}{X_f}\left(\epsilon - \frac{1}{3}\epsilon^2 + \frac{7}{45}\epsilon^3\right) + O(\epsilon^4) \tag{2.105}$$

Integrating Eq. (2.105) and imposing the initial condition $\tau = 0$, $X_f = 0$, we get

$$X_f^2 = 2\tau\left(\epsilon - \frac{1}{3}\epsilon^2 + \frac{7}{45}\epsilon^3\right) + O(\epsilon^4) \tag{2.106}$$

If we wish to express τ as a function of X_f, we write

$$\tau = \frac{1}{2}X_f^2\epsilon^{-1}\left(1 - \frac{1}{3}\epsilon + \frac{7}{45}\epsilon^2\right)^{-1}$$

Using the binomial expansion,

$$\left(1 - \frac{1}{3}\epsilon + \frac{7}{45}\epsilon^2\right)^{-1} = 1 + \frac{1}{3}\epsilon - \frac{2}{45}\epsilon^2 + O(\epsilon^3)$$

Thus,

$$\tau = \frac{1}{2}\epsilon^{-1}X_f^2 + \frac{1}{6}X_f^2 - \frac{1}{45}\epsilon X_f^2 + O(\epsilon^2) \tag{2.107}$$

By using the similarity technique it is possible to obtain an exact solution of Eqs. (1.23)–(1.25) as (Carslaw and Jaeger, 1959)

$$\theta = 1 - \frac{\operatorname{erf}(\lambda X/X_f)}{\operatorname{erf}(\lambda)} \tag{2.108}$$

where λ is a root of the equation

$$\sqrt{\pi}\,\lambda e^{\lambda^2} \operatorname{erf}(\lambda) = \epsilon \tag{2.109}$$

and

$$X_f = 2\lambda\tau^{1/2} \tag{2.110}$$

In Fig. 2.5, the exact solution for the freezing time obtained from Eqs. (2.109) and (2.110) is compared with the perturbation solution in Eq. (2.107) for $\epsilon = 0.2, 0.4$, and 0.6.

2.8 TWO-DIMENSIONAL STEADY CONDUCTION IN A BODY OF IRREGULAR SHAPE

The equations for this example were derived in Section 1.2.5. Here we derive a two-term regular perturbation solution for Eqs. (1.34) and (1.35) by assuming

$$\theta = \theta_0(R) + \epsilon\theta_1(R, \psi) + O(\epsilon^2) \tag{2.111}$$

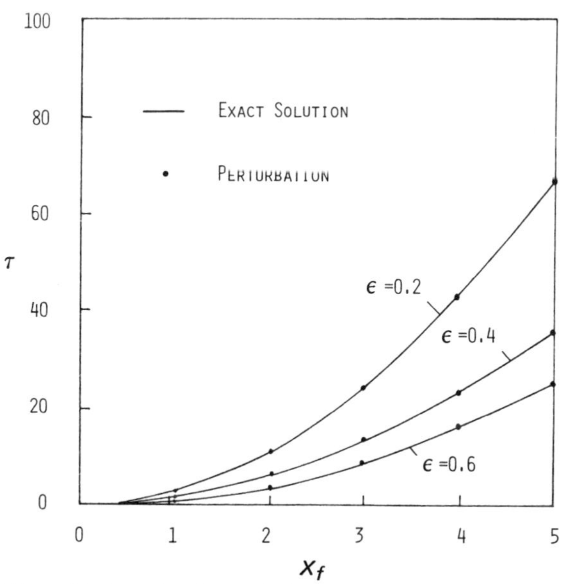

Figure 2.5 Freezing time solution.

Following the usual procedure, the governing equations for θ_0 and θ_1 can be derived as

$$\epsilon^0: \quad \frac{d^2\theta_0}{dR^2} + \frac{1}{R+h}\frac{d\theta_0}{dR} = 0 \tag{2.112}$$

$$\epsilon^1: \quad \frac{\partial^2\theta_1}{\partial R^2} + \frac{1}{R+h}\frac{\partial\theta_1}{\partial R} + \frac{1}{(R+h)^2}\frac{\partial^2\theta_1}{\partial\psi^2} = 0 \tag{2.113}$$

Next, let us look at the first boundary condition in Eq. (1.35) which readily gives

$$R = 0 \qquad \theta_0(0) = 1 \tag{2.114}$$

$$R = 0 \qquad \theta_1(0) = 0 \tag{2.115}$$

However, the second boundary condition in Eq. (1.35) appears as

$$\theta_0[R = 1 + \epsilon f(\psi)] + \epsilon\theta_1[R = 1 + \epsilon f(\psi), \psi] = 0 \tag{2.116}$$

In Eq. (2.116) the perturbation quantity ϵ appears implicitly as well as explicitly. The implicitness is associated with the variable $R = 1 + \epsilon f(\psi)$. In order to compare the coefficients of powers of ϵ, it is necessary to express Eq. (2.116) explicitly in terms of ϵ. This can be done by expanding θ_0 and θ_1 in Taylor series about $R = 1$. Retaining the first two terms of the series gives

$$\theta_0(1) + \epsilon[\theta_1(1, \psi) + f(\psi)\theta_0'(1)] + O(\epsilon^2) = 0 \tag{2.117}$$

where ϵ now appears only explicitly. The prime denotes differentiation with respect to R. Equating coefficients of ϵ^0 and ϵ^1, it follows from Eq. (2.117) that

$$R = 1 \qquad \theta_0(1) = 0 \tag{2.118}$$

$$R = 1 \qquad \theta_1(1, \psi) = -f(\psi)\theta_0'(1) \tag{2.119}$$

The solution of Eq. (2.112) subject to Eqs. (2.114) and (2.118) corresponds to one-dimensional conduction in a hollow cylinder and is well known. The problem described by Eq. (2.113) subject to Eqs. (2.115) and (2.119) now constitutes the standard problem of two-dimensional conduction in a regular hollow cylinder and its solution can be obtained using the method of separation of variables (Scheffler, 1974). Instead of pursuing this problem further, we retreat to a situation which permits a simpler solution for θ_1. Assume that the outer surface (see Fig. 1.5) can be approximated as

$$r = r_2 - a\cos\psi \tag{2.120}$$

so that

$$R = 1 - \epsilon\cos\psi \tag{2.121}$$

Next assume an expansion for θ of the form

$$\theta = \theta_0(R) + (\epsilon \cos \psi)[\theta_1(R)] + O(\epsilon^2) \tag{2.122}$$

With Eq. (2.122), the governing equations for θ_0 remain unchanged but θ_1 is now described by the following ordinary differential equation

$$\frac{d^2\theta_1}{dR^2} + \frac{1}{R+h}\frac{d\theta_1}{dR} - \frac{1}{(R+h)^2}\theta_1 = 0 \tag{2.123}$$

The boundary condition in Eq. (2.117) now takes the form

$$\theta_0(1) + \epsilon \cos \psi [\theta_1(1) - \theta_0'(1)] + O(\epsilon^2) = 0 \tag{2.124}$$

which gives

$$R = 1 \qquad \theta_0(1) = 0 \tag{2.125}$$

$$R = 1 \qquad \theta_1(1) = \theta_0'(1) \tag{2.126}$$

As remarked earlier, the solution for θ_0 is readily obtained as

$$\theta_0 = 1 - \frac{\ln[(R+h)/h]}{\ln[(1+h)/h]} \tag{2.127}$$

Equation (2.123) is recognized as the Euler equation whose solution is

$$\theta_1 = C(R+h) + \frac{D}{R+h}$$

where C and D are constants. Applying the boundary conditions in Eqs. (2.125) and (2.126), C and D are evaluated as

$$C = -\frac{1}{(1+2h)\ln[(1+h)/h]} \qquad \text{and} \qquad D = \frac{h^2}{(1+2h)\ln[(1+h)/h]}$$

Thus

$$\theta_1 = \frac{1}{(1+2h)\ln[(1+h)/h]}\left[\frac{h^2}{R+h} - (R+h)\right] \tag{2.128}$$

2.9 LAMINAR NATURAL CONVECTION FROM A THIN VERTICAL CYLINDER

With reference to the formulation in Section 1.2.6 we now propose to solve Eqs. (1.44)-(1.46) by assuming expansions for F and θ of the form

$$F = \sum_{n=0}^{\infty} \epsilon^n F_n \tag{2.129}$$

$$\theta = \sum_{n=0}^{\infty} \epsilon^n \theta_n \tag{2.130}$$

Since derivatives up to third order are involved, we differentiate Eqs. (2.129) and (2.130) with respect to η and obtain for the mth derivative

$$F^{(m)} = \sum_{n=0}^{\infty} \epsilon^n F_n^{(m)} \tag{2.131}$$

$$\theta^{(m)} = \sum_{n=0}^{\infty} \epsilon^n \theta_n^{(m)} \tag{2.132}$$

Equations (1.44) and (1.45) also involve derivatives with respect to ϵ. Differentiating Eqs. (2.131) and (2.132) with respect to ϵ, we have

$$\frac{\partial F^{(m)}}{\partial \epsilon} = \sum_{n=0}^{\infty} n\epsilon^{n-1} F_n^{(m)} \tag{2.133}$$

$$\frac{\partial \theta^{(m)}}{\partial \epsilon} = \sum_{n=0}^{\infty} n\epsilon^{n-1} \theta_n^{(m)} \tag{2.134}$$

Making use of Eqs. (2.129)-(2.134) in Eqs. (1.44)-(1.46), equating coefficients of like powers of ϵ, and truncating the expansion at the third term, we get

$$\epsilon^0: \quad F_0''' + 3F_0F_0'' - 2(F_0')^2 + \theta_0 = 0 \tag{2.135}$$

$$\theta_0'' + 3\,\mathrm{Pr}\,F_0\theta_0' = 0 \tag{2.136}$$

$$\epsilon^1: \quad F_1''' + 3F_0F_1'' - 5F_0'F_1' + 4F_0''F_1 + \theta_1 + \eta F_0''' + F_0'' = 0 \tag{2.137}$$

$$\theta_1'' + 3\,\mathrm{Pr}\,F_0\theta_1' - \mathrm{Pr}\,F_0'\theta_1 + 4\,\mathrm{Pr}\,F_1\theta_0' + \eta\theta_0'' + \theta_0' = 0 \tag{2.138}$$

$$\epsilon^2: \quad F_2''' + 3F_0F_2'' - 6F_0'F_2' + 5F_0''F_2 + \theta_2 + \eta F_1''' + F_1'' + 4F_1F_1'' - 3(F_1')^2 = 0 \tag{2.139}$$

$$\theta_2'' + 3\,\mathrm{Pr}\,F_0\theta_2' - 2\,\mathrm{Pr}\,F_0'\theta_2 + 5\,\mathrm{Pr}\,\theta_0'F_2 + \eta\theta_1'' + \theta_1' + 4\,\mathrm{Pr}\,F_1\theta_1' - \mathrm{Pr}\,\theta_1F_1' = 0 \tag{2.140}$$

The boundary conditions are

$$\eta = 0 \qquad F_0 = F_1 = F_2 = 0 \qquad F_0' = F_1' = F_2' = 0 \tag{2.141}$$

$$\eta = 0 \qquad \theta_0 = 1 \qquad \theta_1 = \theta_2 = 0 \tag{2.142}$$

$$\eta = \infty \qquad F_0' = F_1' = F_2' = 0 \tag{2.143}$$

$$\eta = \infty \qquad \theta_0 = \theta_1 = \theta_2 = 0 \tag{2.144}$$

The zero-order approximation given by Eqs. (2.135) and (2.136) corresponds to a vertical plate and is nonlinear. However, the subsequent-order problems, Eqs. (2.137)-(2.140), which give corrections for the transverse curvature effect, are linear.

Unlike the previous examples, we cannot obtain analytical solutions. The only feasible approach is to use a numerical scheme and here a variety of choices is available (Na, 1979). We quote the results of one such scheme. For Pr = 0.72, the solution for $F''(0)$, which is proportional to the skin friction, is

$$F''(0) = 0.6760 + 0.0595\epsilon - 0.0038\epsilon^2 \tag{2.145}$$

The corresponding result for $\theta'(0)$, which gives the heat transfer rate, is

$$-\theta'(0) = 0.5046 + 0.2316\epsilon - 0.0307\epsilon^2 \tag{2.146}$$

In Table 2.2 we compare the present results for $F''(0)$ and $\theta'(0)$ with the local nonsimilarity solutions reported by Minkowycz and Sparrow (1974). Up to $\epsilon = 1$ the present solution agrees fairly well with the local nonsimilarity solution, but as ϵ increases, the accuracy deteriorates.

2.10 UNSTEADY HEAT TRANSFER FOR LAMINAR FLOW OVER A FLAT PLATE

The momentum and energy equations for this problem were derived in Section 1.2.7. As they stand, they are appropriate for studying the boundary-

Table 2.2 Values of $F''(0)$ and $-\theta'(0)$ for Pr = 0.72

	$F''(0)$		$-\theta'(0)$	
ϵ	Perturbation solution	Local nonsimilarity	Perturbation solution	Local nonsimilarity
0	0.6760	0.6741	0.5046	0.5079
0.1	0.6819	0.6800	0.5274	0.5302
0.2	0.6877	0.6861	0.5497	0.5526
0.5	0.7048	0.7037	0.6127	0.6156
0.75	0.7185	0.7178	0.6610	0.6660
1	0.7317	0.7316	0.7055	0.7138
5	0.8785	0.9195	0.8951	1.354
10	0.8910	1.117	−0.2494	2.016

layer behavior for large time, that is, the approach to the final steady state. During the early stages of thermal boundary-layer development, the dominant process must be conduction, and hence it seems appropriate to introduce a new similarity variable ξ of the form used in pure conduction problems, that is,

$$\xi = \frac{y}{2(\alpha t)^{1/2}} = \frac{\eta}{2}\frac{\mathrm{Pr}}{\epsilon} \tag{2.147}$$

Writing Eqs. (1.55) and (1.56) in terms of ξ gives

$$\theta'' + [2\xi + (\epsilon\,\mathrm{Pr})^{1/2}(f - \xi f')]\,\theta' = (4\epsilon - 2\,\mathrm{Pr}^{1/2}\,\epsilon^{3/2} f')\frac{\partial\theta}{\partial\epsilon} \tag{2.148}$$

$$\theta(0, \epsilon) = 1 \qquad \theta(\infty, \epsilon) = 0 \tag{2.149}$$

$$\epsilon = 0 \qquad \theta(\xi, \epsilon) = 0 \tag{2.150}$$

where primes now denote differentiation with respect to ξ.

At small values of time, the thickness of the thermal boundary-layer is much smaller compared to the thickness of the velocity boundary-layer. Hence, one may approximate the Blasius function f by Welyl's expansion near the surface

$$f = \sum_{n=0}^{\infty} \beta\alpha^{n+1} D_n \left(-\frac{\eta}{2}\right)^{2+3n} \qquad D_0 = 2 \tag{2.151}$$

$$(3n+2)(3n+1)(3n)D_n = \sum_{i=0}^{n-1} (3i+2)(3i+1)D_i D_{n-1-i} \tag{2.152}$$

where $\beta = f''(0) = 0.33206$. Using only two terms of the expansion in Eq. (2.151)

$$f = \frac{1}{2}\beta\eta^2 - \frac{1}{240}\beta^2\eta^5 \tag{2.153}$$

Changing the variable to ξ in Eq. (2.153) and substituting for f in Eq. (2.148), the energy equation now becomes

$$\theta'' + \left(2\xi - \frac{2\beta}{\mathrm{Pr}^{1/2}}\xi^2\epsilon^{3/2} + \frac{8\beta^2}{15\,\mathrm{Pr}^2}\xi^5\epsilon^3\right)\theta' = \left(4\epsilon - \frac{8\beta}{\mathrm{Pr}^{1/2}}\xi\epsilon^{5/2} + \frac{4}{3}\frac{\beta^2}{\mathrm{Pr}^2}\xi^4\epsilon^4\right)\frac{\partial\theta}{\partial\epsilon} \tag{2.154}$$

Riley (1963) has shown that the expansion for Eq. (2.154) must take the form

$$\theta = \sum_{n=0}^{\infty} \epsilon^{(3/2)n}\, \theta_n \tag{2.155}$$

Here we restrict ourselves to the first two terms so that

$$\theta = \theta_0 + \epsilon^{3/2}\, \theta_1 \tag{2.156}$$

The governing equations for θ_0 and θ_1 then follow as

$$\epsilon^0: \quad \theta_0'' + 2\xi\,\theta_0' = 0 \tag{2.157}$$

$$\theta_0(0) = 1 \qquad \theta_0(\infty) = 0 \tag{2.158}$$

$$\epsilon^{3/2}: \quad \theta_1'' + 2\xi\,\theta_1' - 6\theta_1 = \frac{2\beta}{\mathrm{Pr}^{1/2}}\,\xi^2\,\theta_0' \tag{2.159}$$

$$\theta_1(0) = 0 \qquad \theta_1(\infty) = 0 \tag{2.160}$$

The zero-order problem in Eqs. (2.157) and (2.158) represents the classical problem of transient conduction into a semi-infinite medium and its solution is

$$\theta_0 = \operatorname{erfc}\xi = \frac{2}{\sqrt{\pi}} \int_{\xi}^{\infty} e^{-u^2}\, du \tag{2.161}$$

where erfc stands for the complimentary error function. The $\frac{3}{2}$-order problem in Eq. (2.159) now becomes

$$\theta_1'' + 2\xi\,\theta_1' - 6\theta_1 = -\frac{4\beta}{(\pi\,\mathrm{Pr})^{1/2}}\,\xi^2 e^{-\xi^2} \tag{2.162}$$

The solution of Eq. (2.162) may be assumed to be of the form (Van Dyke, 1975b)

$$\theta_1 = C_1 \xi^2 e^{-\xi^2} + C_2 \xi \operatorname{erfc}\xi + C_3 \xi^3 \operatorname{erfc}\xi \tag{2.163}$$

Substituting Eq. (2.163) into Eq. (2.162) and equating coefficients of like functions of ξ, the solutions for C_1, C_2, and C_3 are obtained as

$$C_1 = \frac{\beta}{4\sqrt{\pi\,\mathrm{Pr}}} \qquad C_2 = \frac{\beta}{8\sqrt{\mathrm{Pr}}} \qquad C_3 = \frac{\beta}{12\,\mathrm{Pr}^{1/2}} \tag{2.164}$$

Thus the solution for θ_1 is

$$\theta_1 = \frac{\beta}{12\sqrt{\pi\,\mathrm{Pr}}} \left[3\xi^2 e^{-\xi^2} + \frac{\sqrt{\pi}}{2}(3\xi + 2\xi^3) \operatorname{erfc}\xi \right] \tag{2.165}$$

The series for the rate of heat transfer per unit area, q, can now be derived. In dimensionless form, it is

$$Q = \frac{qx}{k(T_0 - T_\infty)\,\mathrm{Re}^{1/2}} = -\frac{1}{2}\left(\frac{\mathrm{Pr}}{\epsilon}\right)^{1/2}\theta'(0) = \left(\frac{\mathrm{Pr}}{\pi\epsilon}\right)^{1/2} - \frac{\beta}{16}\epsilon$$

$$= 0.5642\,\mathrm{Pr}^{1/2}\,\epsilon^{-1/2} - 0.0208\beta\epsilon \tag{2.166}$$

where $\mathrm{Re} = U_\infty x/\nu$ is the Reynolds number. The perturbation solution for Q is compared with the numerical solution (Ingham, 1977) in Table 2.3 for $\mathrm{Pr} = 0.1$, 1.0, and 10. The values appearing in the parentheses are from the numerical solution. The comparison is favorable.

2.11 ENTRANCE REGION HEAT TRANSFER IN A TUBE

Consider the steady flow of an incompressible fluid through a circular tube of radius r_0 whose surface is maintained at T_w (see Fig. 2.6). The fluid entering the tube is assumed to be at a uniform temperature T_i. Let u be the axial velocity of the fluid. The energy equation describing the thermal development of flow is

$$\frac{1}{r}\frac{\partial}{\partial r}\left(r\frac{\partial T}{\partial r}\right) = \frac{u}{\alpha}\frac{\partial T}{\partial z} \tag{2.167}$$

$$z = 0 \qquad T = T_i \tag{2.168}$$

$$r = r_0 \qquad T = T_w \tag{2.169}$$

$$r = 0 \qquad \frac{\partial T}{\partial r} = 0 \tag{2.170}$$

Table 2.3 Values of Q at different Pr

	Q		
ϵ	Pr = 0.1	Pr = 1.0	Pr = 10
0.10	0.5621 (0.5621)	1.7820 (1.7819)	5.6398 (5.6419)
0.25	0.3516 (0.3518)	1.1232 (1.1228)	3.5630 (3.5640)
0.50	0.2419 (0.2427)	0.7875 (0.7866)	2.5127 (2.5226)
0.75	0.1904 (0.1917)	0.6359 (0.6338)	2.0445 (2.0435)
1.00	0.1577 (0.1586)	0.5434 (0.5392)	1.7633 (1.7610)

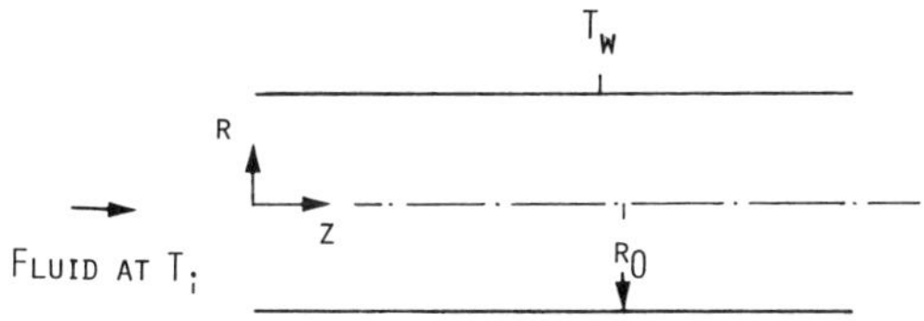

Figure 2.6 Entrance region heat transfer in a tube.

where r, z, and q are the radial coordinate, axial coordinate, and thermal diffusivity respectively. In writing Eq. (2.167), it is assumed that axial conduction and viscous dissipation are negligible. It is further assumed that the flow is hydrodynamically developed so that

$$u = 2V\left(1 - \frac{r^2}{r_0^2}\right) \tag{2.171}$$

where V is the mean velocity.

Introducing the following dimensionless quantities

$$\theta = \frac{T_w - T}{T_w - T_i} \qquad R = \frac{r}{r_0} \qquad Z = \frac{\alpha z}{2r_0^2 V} \tag{2.172}$$

Equations (2.167)-(2.170) take the form

$$\frac{1}{R}\frac{\partial}{\partial R}\left(R\frac{\partial\theta}{\partial R}\right) = (1 - R^2)\frac{\partial\theta}{\partial Z} \tag{2.173}$$

$$\theta(R,0) = 1 \qquad \theta(1,Z) = 0 \qquad \frac{\partial\theta}{\partial R}(0,Z) = 0 \tag{2.174}$$

Unlike the previous examples, we have not so far introduced the perturbation quantity ϵ. To identify ϵ we consider the Leveque solution which is applicable if the curvature effect is neglected and a linear velocity profile is assumed. In this case, the introduction of a similarity variable $\eta = (1 - R) \times (2/9Z)^{1/3}$ reduces the partial differential equation to an ordinary differential equation. This motivates us to introduce two parameters η and ϵ such that

$$\eta = (1 - R)\left(\frac{2}{9Z}\right)^{1/3} \qquad \epsilon = \left(\frac{9Z}{2}\right)^{1/3} \tag{2.175}$$

Introducing Eq. (2.175), Eqs. (2.173)-(2.174) now turn into

$$(1 - \epsilon\eta)\epsilon^{-2}\theta'' - \epsilon^{-1}\theta' = [1 - (1 - \epsilon\eta)^2](1 - \epsilon\eta)\left(\frac{3}{2}\epsilon^{-2}\frac{\partial\theta}{\partial\epsilon} - \frac{3}{2}\eta\theta'\right) \tag{2.176}$$

$$\theta(\infty,\epsilon) = 1 \qquad \theta(0,\epsilon) = 0 \tag{2.177}$$

where primes denote differentiation with respect to η.

It should be noted that the problem now has the character of a boundary-layer and the symmetry condition $\partial\theta/\partial R(0, Z) = 0$ is not applied.

Let us derive a three-term perturbation solution by assuming

$$\theta = \theta_0 + \epsilon\theta_1 + \epsilon^2\theta_2 \tag{2.178}$$

Then the governing equations for θ_0, θ_1, and θ_2 are

$$\theta_0'' + 3\eta^2\theta_0' = 0 \tag{2.179}$$

$$\theta_0(\infty) = 1 \qquad \theta_0(0) = 0 \tag{2.180}$$

$$\theta_1'' + 3\eta^2\theta_1' - 3\eta\theta_1 = 1 + \tfrac{3}{2}\eta^3\theta_0' \tag{2.181}$$

$$\theta_1(\infty) = 0 \qquad \theta_1(0) = 0 \tag{2.182}$$

$$\theta_2'' + 3\eta^2\theta_2' - 6\eta\theta_2 = (1 + \tfrac{3}{2}\eta^3)\theta_1' - \tfrac{3}{2}\eta^2\theta_1 + \eta\theta_0' \tag{2.183}$$

$$\theta_2(\infty) = 0 \qquad \theta_2(0) = 0 \tag{2.184}$$

Shih and Tsou (1978) have solved Eqs. (2.179)-(2.184) numerically. Using their results, one can obtain $\theta_0'(0)$, $\theta_1'(0)$, and $\theta_2'(0)$ which relate to the local Nusselt number Nu_Z as

$$\mathrm{Nu}_Z = \frac{2}{\epsilon}\,\theta'(0) = \frac{2}{\epsilon}\,(1.11985 - 0.60000\epsilon - 0.08982\epsilon^2) \tag{2.185}$$

The average Nusselt number over a distance Z follows from Eq. (2.185) as

$$\mathrm{Nu} = 1.75053\,\mathrm{Gz}^{1/3} - 1.2000 - 0.25887\,\mathrm{Gz}^{-1/3} \tag{2.186}$$

where $\mathrm{Gz} = \pi r_0^2 u\rho c/kz$ is the well-known Graetz number.

2.12 CONVECTING–RADIATING FIN WITH VARIABLE THERMAL CONDUCTIVITY

Here we propose to use a double series expansion to solve Eqs. (1.68) and (1.69). As mentioned earlier, ϵ_1 is usually small, and for the case of weak radiation-conduction interaction, ϵ_2 is also small. Therefore, an asymptotic expansion for θ in ϵ_1 and ϵ_2 may be assumed to be

$$\theta = \sum_{j=0}^{\infty}\sum_{i=0}^{j} \epsilon_1^{j-i}\epsilon_2^{i}\theta_{i,j-1} \tag{2.187}$$

The case of $\theta_a = \theta_s = 0$ is considered. The solution for the more general case can be obtained in a similar manner though the algebra is somewhat lengthy. Assuming second-order expansion,

$$\theta = \theta_{00} + \epsilon_1\theta_{01} + \epsilon_2\theta_{10} + \epsilon_1^2\theta_{02} + \epsilon_1\epsilon_2\theta_{11} + \epsilon_2^2\theta_{20} \tag{2.188}$$

Substituting Eq. (2.188) into Eq. (1.68) and taking $\theta_a = \theta_s = 0$, we obtain the following system of equations

$$\frac{d^2\theta_{00}}{dX^2} - N^2\theta_{00} = 0 \tag{2.189}$$

$$\frac{d^2\theta_{01}}{dX^2} - N^2\theta_{01} = -\left(\frac{d\theta_{00}}{dX}\right)^2 - \theta_{00}\frac{d^2\theta_{00}}{dX^2} \tag{2.190}$$

$$\frac{d^2\theta_{10}}{dX^2} - N^2\theta_{10} = \theta_{00}^4 \tag{2.191}$$

$$\frac{d^2\theta_{02}}{dX^2} - N^2\theta_{02} = -\theta_{00}\frac{d^2\theta_{01}}{dX^2} - \theta_{01}\frac{d^2\theta_{00}}{dX^2} - 2\frac{d\theta_{00}}{dX}\frac{d\theta_{01}}{dX} \tag{2.192}$$

$$\frac{d^2\theta_{11}}{dX^2} - N^2\theta_{11} = 4\theta_{00}^3\theta_{01} - \theta_{10}\frac{d^2\theta_{00}}{dX^2} - \theta_{00}\frac{d^2\theta_{10}}{dX^2} - 2\frac{d\theta_{00}}{dX}\frac{d\theta_{10}}{dX} \tag{2.193}$$

$$\frac{d^2\theta_{20}}{dX^2} - N^2\theta_{20} = 4\theta_{00}^3\theta_{10} \tag{2.194}$$

The boundary conditions are

$$\begin{aligned} X = 0 \quad & \frac{d\theta_{00}}{dX} = \frac{d\theta_{01}}{dX} = \frac{d\theta_{10}}{dX} = \frac{d\theta_{02}}{dX} = \frac{d\theta_{11}}{dX} = \frac{d\theta_{20}}{dX} = 0 \\ x = 1 \quad & \theta_{00} = 1 \quad \theta_{01} = \theta_{10} = \theta_{02} = \theta_{11} = \theta_{20} = 0 \end{aligned} \tag{2.195}$$

The solution for θ_{00} can be obtained in a straightforward manner. By using the method of variation of parameters, we can solve the subsequent equations. The final solutions are

$$\theta_{00} = \operatorname{sech} N \cosh NX \tag{2.196}$$

$$\theta_{01} = \frac{1}{3}(\operatorname{sech}^2 N)(\cosh 2N \operatorname{sech} N \cosh NX - \cosh 2NX) \tag{2.197}$$

$$\begin{aligned} \theta_{10} = \frac{3}{8}\frac{\operatorname{sech}^4 N}{N^2}\Bigg[&\left(1 - \frac{4}{9}\cosh 2N - \frac{1}{45}\cosh 4N\right)\operatorname{sech} N \cosh NX \\ &+ \frac{4}{9}\cosh 2NX + \frac{1}{45}\cosh 4NX - 1\Bigg] \end{aligned} \tag{2.198}$$

$$\begin{aligned} \theta_{02} = \frac{1}{6}(\operatorname{sech}^3 N)\Bigg[&\Bigg(\frac{4}{3}\operatorname{sech}^2 N \cosh^2 2N - \frac{9}{8}\operatorname{sech} N \cosh 3N \\ &- \frac{1}{2}N \tanh N\Bigg)\cosh NX - \frac{4}{3}\operatorname{sech} N \cosh 2N \cosh 2NX \\ &+ \frac{9}{8}\cosh 3NX + \frac{1}{2}NX \sinh NX\Bigg] \end{aligned} \tag{2.199}$$

$$
\begin{aligned}
\theta_{20} = \frac{3}{4}\frac{\operatorname{sech}^7 N}{N^4} \Bigg\{ & \left(\frac{1}{3}\cosh 2N + \frac{1}{60}\cosh 4N - \frac{3}{4}\right)\operatorname{sech} N \\
& + \Bigg[\left(\frac{3}{4} - \frac{2}{3}\cosh 2N - \frac{1}{30}\cosh 4N + \frac{2}{135}\cosh 2N \cosh 4N\right. \\
& \left. + \frac{4}{27}\cosh^2 2N + \frac{1}{2700}\cosh^2 4N\right)\operatorname{sech} N + \frac{3}{160}\cosh 3N \\
& - \frac{23}{4320}\cosh 5N - \frac{1}{8640}\cosh 7N + \frac{21}{40} N \sin N \operatorname{sech} N\Bigg]\cosh NX \\
& - \left(\frac{4}{27}\cosh 2N + \frac{1}{135}\cosh 4N - \frac{1}{3}\right)\operatorname{sech} N \cosh 2NX \\
& - \frac{3}{160}\cosh 3NX - \left(\frac{1}{135}\cosh 2N + \frac{1}{2700}\cosh 4N - \frac{1}{60}\right) \\
& \times \operatorname{sech} N \cosh 4NX + \frac{23}{4320}\cosh 5NX + \frac{1}{8640}\cosh 7NX \\
& - \frac{21}{40} NX \sinh NX \Bigg\}
\end{aligned} \tag{2.200}
$$

In Fig. 2.7 the perturbation solutions for $\epsilon_2 = 0.2$ and 0.6 for various values of parameter ϵ_1 are compared with the corresponding numerical solutions, the parameter N being fixed at unity. For $\epsilon_2 = 0.2$, the perturbation solutions match almost exactly with numerical solutions except for slight deviation for $\epsilon_1 = 0.6$. As ϵ_2 increases, the accuracy of the perturbation solutions decreases particularly at high values of ϵ_1. But even at $\epsilon_1 = 0.6$, where the maximum discrepancy occurs, the perturbation tip temperature is still within 2 percent of the numerical solution. Further calculations show that the accuracy of the perturbation solution increases as N increases and decreases as N is reduced. Sparrow and Niewerth (1968) have found that for optimum operating conditions, ϵ_2 (N_r in their notation) varies from 0.7 to 0 as N (N_{cv} in their notation) varies from 0 to 1.4192. Hence, it is obvious that the perturbation solution would yield accurate temperature distribution (and hence heat flux and fin efficiency) over the major portion of the range of optimum operating conditions.

2.13 CONVECTING–RADIATING COOLING OF A LUMPED SYSTEM WITH VARIABLE SPECIFIC HEAT

As another example of double series expansion we consider the problem of combined convective-radiative cooling of a lumped system with temperature-

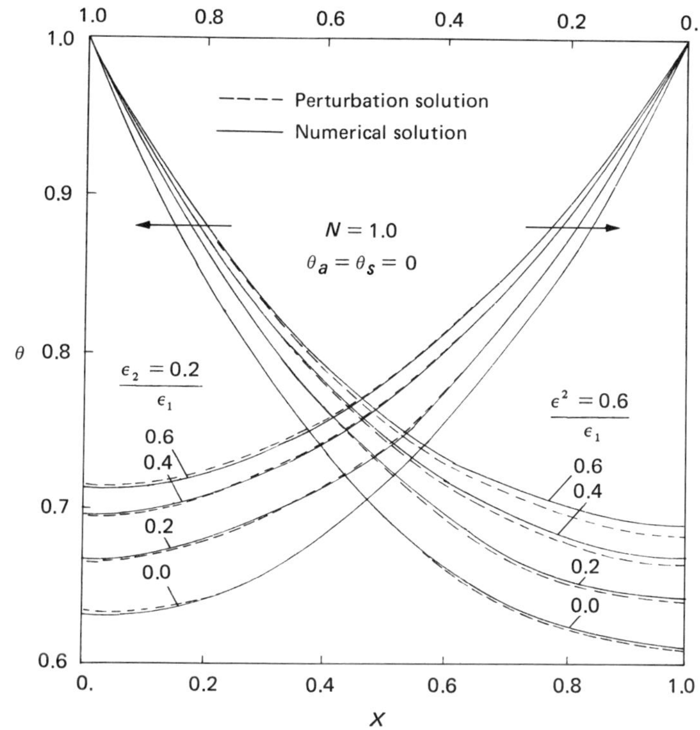

Figure 2.7 Temperature distribution in a convecting-radiating fin.

dependent specific heat (Aziz and Hamad, 1977). Let the system have volume V, surface area A, density ρ, specific heat C, emissivity E, and initial temperature T_i. At time $t = 0$, the system is exposed to an environment with convective heat transfer coefficient h and temperature T_a. The system also loses heat by radiation, the effective sink temperature being T_s. Assume that the specific heat C is a linear function of temperature of the form

$$C = C_a[1 + \beta(T - T_a)] \tag{2.201}$$

where C_a is the specific heat at temperature T_a and β is a constant. The cooling equation and the initial condition are

$$\rho V C \frac{dT}{dt} + hA(T - T_a) + E\sigma A(T^4 - T_s^4) = 0 \tag{2.202}$$

$$t = 0 \qquad T = T_i \tag{2.203}$$

Introducing Eq. (2.201) and using the dimensionless parameters

$$\theta = \frac{T}{T_i} \qquad \theta_a = \frac{T_a}{T_i} \qquad \theta_s = \frac{T_s}{T_i} \qquad \tau = \frac{t}{\rho V C_a / hA} \qquad \epsilon_1 = \beta T_i \qquad \epsilon_2 = \frac{E\sigma T_i^3}{h} \tag{2.204}$$

transforms Eqs. (2.202) and (2.203) to

$$[1 + \epsilon_1(\theta - \theta_a)] \frac{d\theta}{d\tau} + (\theta - \theta_a) + \epsilon_2(\theta^4 - \theta_s^4) = 0 \tag{2.205}$$

$$\tau = 0 \qquad \theta = 1 \tag{2.206}$$

For simplicity, we consider the case of $\theta_a = \theta_s = 0$. Assuming an expansion of the form in Eq. (2.188) and substituting it into Eq. (2.205), we obtain the following system of equations

$$\frac{d\theta_{00}}{d\tau} + \theta_{00} + 0 \tag{2.207}$$

$$\frac{d\theta_{01}}{d\tau} + \theta_{01} + \theta_{00} \frac{d\theta_{00}}{d\tau} = 0 \tag{2.208}$$

$$\frac{d\theta_{10}}{d\tau} + \theta_{10} + \theta_{00}^4 = 0 \tag{2.209}$$

$$\frac{d\theta_{02}}{d\tau} + \theta_{11} + \theta_{00} \frac{d\theta_{01}}{d\tau} + \theta_{01} \frac{d\theta_{00}}{d\tau} = 0 \tag{2.210}$$

$$\frac{d\theta_{11}}{d\tau} + \theta_{11} + \theta_{00} \frac{d\theta_{10}}{d\tau} + \theta_{10} \frac{d\theta_{00}}{d\tau} + 4\theta_{00}^3\theta_{01} = 0 \tag{2.211}$$

$$\frac{d\theta_{20}}{d\tau} + \theta_{20} + 4\theta_{00}^3\theta_{10} = 0 \tag{2.212}$$

The initial conditions are

$$\tau = 0 \qquad \theta_{00} = 1 \qquad \theta_{01} = \theta_{02} = \theta_{10} = \theta_{20} = \theta_{11} = 0 \tag{2.213}$$

The solutions of Eqs. (2.207)-(2.212) subject to Eq. (2.213) are easily obtained. The final solution correct to second-order is

$$\theta = e^{-\tau} + \epsilon_1(e^{-\tau} - e^{-2\tau}) + \frac{1}{3}\epsilon_2(e^{-4\tau} - e^{-\tau}) + \frac{1}{2}\epsilon_1^2(e^{-\tau} - 4e^{-2\tau} + 3e^{-3\tau})$$
$$+ \frac{2}{9}\epsilon_2^2(e^{-\tau} - 2e^{-4\tau} + e^{-7\tau}) + \frac{1}{12}\epsilon_1\epsilon_2(-7e^{-\tau} + 8e^{-2\tau}$$
$$+ 16e^{-4\tau} + 17e^{-5\tau}) \tag{2.214}$$

The accuracy of Eq. (2.214) can be assessed by comparison with the exact solution of Eqs. (2.205) and (2.206) for $\theta_a = \theta_s = 0$. The exact solution can be written as

$$\frac{1}{3}\ln\frac{1+\epsilon_2\theta^3}{(1+\epsilon_2)\theta^3}+\frac{1}{3}\frac{\epsilon_1}{\epsilon_2^{1/3}}\left[\frac{1}{2}\ln\frac{(1+\epsilon_2^{1/3})^3(1+\epsilon_2\theta^3)}{(1+\epsilon_2)(1+\epsilon_2^{1/3}\theta)^3}\right.$$

$$\left.+\sqrt{3}\left(\arctan\frac{2\epsilon^{1/3}-1}{\sqrt{3}}-\arctan\frac{2\epsilon_2^{1/3}\theta-1}{\sqrt{3}}\right)\right]=\tau \quad (2.215)$$

Numerical computations indicate that for ϵ_1, $\epsilon_2 < 0.2$, the perturbation solution in Eq. (2.214) provides a very close approximation to the exact solution in Eq. (2.215).

PROBLEMS

2.1 Adopt a perturbation approach to find the roots of $x^2 - (5+\epsilon)x + 4 = 0$.

2.2 Consider the transcendental equation of Section 2.3. Extend the perturbation solution to eight terms and show that

$$z = \epsilon + \frac{1}{2}\epsilon^3 + \frac{13}{24}\epsilon^5 + \frac{541}{720}\epsilon^7 + \cdots$$

and the error at $\beta = 0.88$, compared to the numerical solution, is now reduced to 18 percent.

2.3 For transient conduction in a slab with convective surface boundary conditions, it is required to find the roots z of the transcendental equation

$$z \tan z = \mathrm{Bi}$$

For $\mathrm{Bi} \gg 1$, carry out a regular perturbation analysis in powers of Bi^{-1} to show that

$$z = z_0 - \mathrm{Bi}^{-1} z_0 + \mathrm{Bi}^{-2} z_0 + \mathrm{Bi}^{-3}\frac{z_0(z_0^2-3)}{3} - \mathrm{Bi}^{-4}\frac{z_0(4z_0^2-3)}{3} + O(\mathrm{Bi}^{-5})$$

where $z_0 = \pi/2, 3\pi/2, 5\pi/2$, etc.

2.4 An analysis for the cooling of a lumped system by combined convection and radiation (Aziz and Hamad, 1977) leads to

$$\frac{d\theta}{d\tau} + (\theta - \theta_a) + \epsilon(\theta^4 - \theta_s^4) = 0$$

$$\tau = 0 \qquad \theta = 1$$

For weak radiation regime ($\epsilon \ll 1$) and $\theta_a = \theta_s = 0$, obtain a regular perturbation solution up to $O(\epsilon^3)$ as

$$\theta = e^{-\tau} + \frac{1}{3}\,\epsilon(e^{-4\tau} - e^{-\tau}) + \frac{2}{9}\,\epsilon^2(e^{-\tau} - 2e^{-4\tau} + e^{-7\tau}) + O(\epsilon^3)$$

Assess the accuracy of the perturbation solution in light of the exact solution

$$\frac{1}{3}\ln\frac{1+\epsilon\theta^3}{(1+\epsilon)\theta^3} = \tau$$

2.5 For a rectangular, purely convecting fin with temperature dependent thermal conductivity (Aziz and Huq, 1975) the temperature distribution equation is

$$(1+\epsilon\theta)\frac{d^2\theta}{dX^2} + \epsilon\left(\frac{d\theta}{dX}\right)^2 - N^2\theta = 0$$

For constant base temperature and insulated tip, the boundary conditions are

$$X = 0 \qquad \frac{d\theta}{dX} = 0$$

$$X = 1 \qquad \theta = 1$$

Show that the three-term perturbation solution is

$$\begin{aligned}\theta = {} & \operatorname{sech} N \cosh NX + \frac{1}{3}\,\epsilon\,\operatorname{sech}^2 N(\cosh 2N \operatorname{sech} N \cosh NX - \cosh 2NX)\\ & + \frac{1}{6}\,\epsilon^2 \operatorname{sech}^3 N\left[\left(\frac{4}{3}\operatorname{sech}^2 N \cosh^2 2N - \frac{9}{8}\operatorname{sech} N \cosh 3N\right.\right.\\ & \left. - \frac{1}{2} N \tanh N\right)\cosh NX - \frac{4}{3}\operatorname{sech} N \cosh 2N \cosh 2NX\\ & \left. + \frac{9}{8}\cosh 3NX + \frac{1}{2} NX \sinh NX\right] + O(\epsilon^3)\end{aligned}$$

2.6 The temperature distribution in a uniformly thick rectangular fin radiating to free space at 0°R is governed by the equation (Aziz, 1979)

$$\frac{d^2\theta}{dX^2} - \epsilon\theta^4 = 0$$

For constant base temperature and insulated tip the boundary conditions are

$$X = 0 \qquad \frac{d\theta}{dX} = 0$$

$$X = 1 \qquad \theta = 1$$

Assume a solution of the form $\theta = \theta_0 + \epsilon\theta_1 + \epsilon^2\theta_2$ and show that

$$\theta_0 = 1 \qquad \theta_1 = \frac{1}{2}(X^2 - 1) \qquad \theta_2 = \frac{1}{6}(X^4 - 6X^2 + 5)$$

2.7 Following the details in Section 1.2.4 formulate the mathematical model describing the freezing of a saturated liquid in semi-infinite region due to convective cooling at $x = 0$ (Huang and Shih, 1975). The first boundary condition in Eq. (1.24) is now

$$k \left.\frac{\partial T}{\partial x}\right|_{x=0} = h[T(0, t) - T_0]$$

where T_0 is the coolant temperature in contact with the wall. Show that a regular perturbation analysis leads to the following two-term solution

$$\theta = \frac{X_f}{1 + X_f}\left(1 - \frac{X}{X_f}\right) + \frac{1}{6}\,\epsilon\,\frac{1}{(1 + X_f)^4}\,[X^2(3 + X)(1 + X_f)$$
$$- X_f^2(3 + X_f)(1 + X)] + O(\epsilon^2)$$

2.8 It has been shown by Pedroso and Domoto (1973d) that, with allowance for variable thermal properties, the boundary value problem governing the freezing of a saturated liquid of semi-infinite extent due to a constant lower wall temperature becomes

$$\theta'' + \epsilon f(\theta)\theta'(1)\eta\theta' = 0$$
$$\theta(0) = 0 \qquad \theta(1) = 1$$

Assuming $f(\theta) = a + b\theta$ and an expansion of the form $\theta = \theta_0 + \epsilon\theta_1 + \epsilon^2\theta_2$, show that the equations governing θ_0, θ_1, and θ_2 are

$$\epsilon^0: \quad \theta_0'' = 0$$
$$\theta_0(0) = 0 \qquad \theta_0(1) = 1$$
$$\epsilon^1: \quad \theta_1'' + \eta(a + b\theta_0)\theta_0'(1)\theta_0' = 0$$
$$\theta_1(0) = 0 \qquad \theta_1(1) = 0$$
$$\epsilon^2: \quad \theta_2'' + \eta\{[\theta_0'(1)\theta_1' + \theta_0'\theta_1'(1)](a + b\theta_0) + b\theta_1\theta_0'(1)\theta_0'\} = 0$$
$$\theta_2(0) = 0 \qquad \theta_2(1) = 0$$

Solve the above sequence of equations to obtain

$$\theta_0 = \eta$$

$$\theta_1 = \frac{\eta}{6}\left[\left(a + \frac{b}{2}\right) - a\eta^2 - \frac{b}{2}\eta^3\right]$$

$$\theta_2 = -\frac{1}{45360}(2394a^2 + 765b^2 + 2772ab)\eta + \frac{1}{36}a(a+b)\eta^3$$

$$+ \frac{1}{144}b^2\eta^4 + \frac{1}{40}a^2\eta^5 + \frac{1}{30}ab\eta^6 + \frac{5}{504}b^2\eta^7$$

2.9 Shih and Ju (1976) consider the inverse Stefan problem of constant speed freezing of a warm liquid in forced convection flow inside or outside a cylinder. For freezing outside the cylinder, the governing equations are

$$\frac{1}{R}\frac{\partial}{\partial R}\left(R\frac{\partial\theta}{\partial R}\right) = \epsilon\frac{\partial\theta}{\partial\tau}$$

$$\theta(R_f, \tau) = 0$$

$$\left.\frac{\partial\theta}{\partial R}\right|_{R=R_f} = -\left(\frac{dR_f}{d\tau} + b\right)$$

$$\left.\frac{\partial\theta}{\partial R}\right|_{R=1} = \text{Bi}\,[\theta(1,\tau) - F(\tau)]$$

where Bi and b are constants. For constant speed freezing assume $R_f = 1 + n\tau$ or $dR_f/d\tau = n$, n being a constant. Assuming $\theta = \theta_0 + \epsilon\theta_1$, show that

$$\theta_0 = (n+b)(1+n\tau)\ln\frac{1+n\tau}{R}$$

$$\theta_1 = \frac{1}{4}n(n+b)\left[2R^2 + R^2\ln\frac{1+n\tau}{R} + (1+n\tau)^2\left(3\ln\frac{1+n\tau}{R} - 2\right)\right]$$

Hence establish that the coolant temperature variation $F(\tau)$ must vary as

$$F(\tau) = (n+b)(1+n\tau)\left[\frac{1}{\text{Bi}} + \ln(1+n\tau)\right] + \frac{1}{4}\epsilon n(n+\text{b})$$

$$\times\left[\left(2 - \frac{3}{\text{Bi}}\right) + \left(1 - \frac{2}{\text{Bi}}\right)\ln(1+n\tau) + \left(\frac{3}{\text{Bi}} - 2\right)(1+n\tau)^2\right.$$

$$\left. + 3(1+n\tau)\ln(1+n\tau)\right]$$

2.10 For transient heat conduction into a semi-infinite medium having temperature dependent heat capacity, the governing equations appearing in Aziz and Benzies (1976) are

$$F'' + 2\eta[1 + \epsilon(F-1)]\,\text{F}' = 0$$

$$\eta = 0 \qquad F = 1$$

$$\eta = \infty \qquad F = 0$$

Obtain a regular perturbation solution as

$$F = \operatorname{erfc} \eta + \epsilon \left[\left(\frac{1}{\pi} + \frac{1}{\sqrt{\pi}} \eta e^{-\eta^2} \right) \operatorname{erf} \eta + \frac{1}{\pi} (e^{-2\eta^2} - 1) \right] + O(\epsilon^2)$$

2.11 In an analysis of steady thermal boundary-layer in liquid metals (Pr $\ll 1$) with variable thermal conductivity, Arunachalam and Rajappa (1978) arrive at the following model

$$(1 + \epsilon H)H'' + 2\eta H' + \epsilon H'^2 = 0$$

$$H(0) = 1 \qquad H(\infty) = 0$$

where the independent variable is η. Assuming that $H = H_0 + \epsilon H_1$, demonstrate that

$$H_0 = \operatorname{erfc} \eta$$

$$H_1 = \left(\frac{1}{2} + \frac{1}{\pi} + \frac{\eta e^{-\eta^2}}{\sqrt{\pi}} \right) \operatorname{erfc} \eta - \frac{1}{2} (\operatorname{erfc} \eta)^2 - \frac{e^{-2\eta^2}}{\pi}$$

2.12 The response of an infinitely long convecting fin to a sudden change in its base temperature is given by (Aziz and Na, 1980)

$$\frac{\partial^2 \theta}{\partial X^2} - N^2 \theta = \frac{\partial \theta}{\partial \tau}$$

$$\theta(0, \tau) = 1 \qquad \theta(X, 0) = \theta(\infty, \tau) = 0$$

Assume a coordinate perturbation expansion of the form

$$\theta = \sum_{n=0}^{\infty} \epsilon^n \theta_n(\eta)$$

where $\eta = X/2\sqrt{\tau}$ and $\epsilon = 4N^2\tau$. Derive the first three perturbation problems and solve them. The solution should appear as

$$\theta = \operatorname{erfc} \eta + \epsilon \left(i^2 \operatorname{erfc} \eta - \frac{1}{4} \operatorname{erfc} \eta \right)$$

$$+ \epsilon^2 \left(i^4 \operatorname{erfc} \eta - \frac{1}{4} i^2 \operatorname{erfc} \eta + \frac{1}{32} \operatorname{erfc} \eta \right) + O(\epsilon^3)$$

where $i^n \operatorname{erfc} \eta$ is the nth repeated integral of the error function.

2.13 In a contribution, Singh (1979) studies the laminar swirling flow in a tube with large suction at the surface and shows that the appropriate equations are

$$F''' + FF'' + \bar{\alpha}\epsilon^2 [G(1 - g_w) + g_w - G^2] = 0$$

$$G'' + FG' = 0$$

$$F(0) = G(\infty) = 1 \qquad F'(0) = G(0) = 0 \qquad F'(\infty) = \epsilon$$

where primes denote differentiation with respect to z. Show that the perturbation analysis gives

$$F = 1 + \epsilon(e^{-z} + z - 1) + O(\epsilon^2)$$

$$G = 1 - e^{-z} + \frac{1}{2}\epsilon(e^{-z} - e^{-2z} + z^2 e^{-z}) + O(\epsilon^2)$$

2.14 Consider the problem of unsteady heat transfer for laminar flow over a flat plate treated in Section 2.10. Show that the governing equations for θ_2 are

$$\epsilon^3: \quad \theta_2'' + 2\xi\theta_2' - 12\theta_2 = \frac{2\beta}{\mathrm{Pr}^{1/2}}(\xi^2\theta_1' - 6\xi\theta_1) - \frac{8\beta^2}{15\,\mathrm{Pr}^2}\xi^5\theta_0'$$

$$\theta_2(0) = 0 \qquad \theta_2(\infty) = 0$$

Solve for θ_2 to obtain

$$\theta_2 = \frac{\beta^2}{6\,\mathrm{Pr}\sqrt{\pi}}\left\{\left[\left(\frac{121\sqrt{\pi}}{160} - \frac{17}{20}\right)\xi + \left(\frac{77\sqrt{\pi}}{120} + \frac{29}{60}\right)\xi^3 + \left(\frac{11\sqrt{\pi}}{120} + \frac{1}{3}\right)\xi^5\right] e^{-\xi^2} - \frac{\sqrt{\pi}}{4}\left(\frac{5\xi^4}{2} + \frac{11\xi^6}{15}\right)\operatorname{erfc}\xi\right\} - \frac{2\beta^2}{15\,\mathrm{Pr}^2\sqrt{\pi}}\left(\frac{\xi}{8} + \frac{\xi^3}{3} + \frac{\xi^5}{3}\right)e^{-\xi^2}$$

Hence show that Q now becomes

$$Q = 0.5642\,\mathrm{Pr}^{1/2}\epsilon^{-1/2} - 0.0208\beta\epsilon - 0.00052\,\mathrm{Pr}^{-3/2}(4.9\,\mathrm{Pr} - 1)\epsilon^{5/2}$$

2.15 A more realistic analysis of plane Couette flow should consider the temperature dependence of both viscosity and thermal conductivity. In this case the governing equations are

$$\frac{d}{dY}\left(\frac{\mu}{\mu_0}\frac{dU}{dY}\right) = 0$$

$$\frac{d}{dY}\left(\frac{k}{k_0}\frac{d\theta}{dY}\right) + \epsilon\frac{\mu}{\mu_0}\left(\frac{dU}{dY}\right)^2 = 0$$

Assuming

$$\frac{k}{k_0} = 1 + \alpha_1\theta + \alpha_2\theta^2$$

$$\frac{\mu_0}{\mu} = 1 + \beta_1 \theta + \beta_2 \theta^2$$

show that

$$\theta = \frac{\epsilon}{2}(Y - Y^2) - \epsilon^2 \frac{\alpha_1}{8}(Y^2 - 2Y^3 + Y^4) - \epsilon^2 \frac{\beta_1}{24}(Y - 2Y^2 + 2Y^3 - Y^4)$$

$$U = Y - \epsilon \frac{\beta_1}{12}(Y - 3Y^2 + 2Y^3) + \epsilon^2 \frac{\beta_1^2}{240}(3Y - 10Y^2 + 10Y^3 - 5Y^4 + 2Y^5)$$

$$+ \frac{\epsilon^2}{240}(\alpha_1 \beta_1 - 2\beta_2)(Y - 10Y^3 + 15Y^4 - 6Y^5)$$

2.16 An alternative approach to solve Eqs. (1.13)–(1.15) is to treat β as the perturbation parameter instead of the Brinkman number. Derive a two-term perturbation solution as

$$U = Y - \frac{1}{12}\beta \,\mathrm{Br}\,(1 - 3Y^2 + 2Y^3) + O(\beta^2)$$

$$\theta = \frac{1}{2}\,\mathrm{Br}\,(Y - Y^2) + \frac{1}{24}\beta \,\mathrm{Br}^2\,(1 - 2Y^3 + Y^4) + O(\beta^3)$$

where Br is the Brinkman number.

CHAPTER

THREE

SINGULAR PERTURBATION EXPANSIONS

3.1 INTRODUCTION

The term *singular perturbation expansion* was briefly mentioned in Section 1.8. The term is used to describe expansions that lack the feature of uniform validity, which characterizes the regular perturbation expansion. In the previous chapter, which was devoted entirely to regular perturbation expansions, the solutions obtained were valid throughout the domain of the independent variable. Now we examine problems in which the expansion fails in certain regions and consequently has little or no usefulness. Such regions of failure are called "regions of nonuniformity" or "boundary-layers." The latter term is reminiscent of the failure of inviscid solution to describe the boundary-layer close to the solid surface.

The nonuniformity exhibits itself in several forms:

1. the solution becomes infinite at some value of independent variable,
2. the solution has a discontinuity within the domain of interest,
3. the solution fails to satisfy some boundary conditions,
4. the solution contains an essential singularity.

In this chapter, we pursue the question of nonuniformity in depth and work a number of problems to highlight the different types of nonuniformities. The deficiency of each solution will be discussed in detail. The remedial measures, if simple, will be offered forthwith but those calling for more elaborate analysis will be deferred to Chapters 4 and 5.

3.2 AN ALGEBRAIC EQUATION

As an elementary example, let us determine the three roots of the cubic equation

$$\epsilon y^3 + y - 2 = 0 \tag{3.1}$$

Assume a regular perturbation expansion of the form

$$y = y_0 + \epsilon y_1 + \epsilon^2 y_2 \tag{3.2}$$

and use it in Eq. (3.1) to obtain the following sequence of algebraic equations

$$\epsilon^0: \quad y_0 = 2 \tag{3.3}$$

$$\epsilon^1: \quad y_1 = -y_0^3 \tag{3.4}$$

$$\epsilon^2: \quad y_2 = -3y_0^2 y_1 \tag{3.5}$$

Thus,

$$y = 2 - 8\epsilon + 96\epsilon^2 \tag{3.6}$$

The expansion in Eq. (3.6) gives only one root, which is real. It fails to give the other two roots, which are complex. The reason is that ϵ multiplies the highest power of y in Eq. (3.1) and the effect of the cubic term is lost in the zero-order solution, Eq. (3.3). Examining the limiting behavior of the roots as $\epsilon \to 0$, it is evident that the other roots tend to infinity, a behavior that is not permitted by the expansion of the form shown in Eq. (3.2). To ensure this behavior, the leading term of the expansion must have the general form

$$y = x\epsilon^{-p} \tag{3.7}$$

where $p > 0$ and is to be determined. Substituting Eq. (3.7) into Eq. (3.1), the first two terms become $x^3\epsilon^{1-3p}$ and $x\epsilon^{-p}$. If we wish to retain the influence of the cubic term, these two terms must balance each other, i.e., $1 - 3p = -p$ or $p = \frac{1}{2}$. Thus, the correct form of expansion for the complex roots is

$$y = x\epsilon^{-1/2} + y_0 + \epsilon^{1/2} y_1 \tag{3.8}$$

Let us now substitute Eq. (3.8) into Eq. (3.1) and retain terms up to $\epsilon^{1/2}$. This gives

$$\epsilon[x^3\epsilon^{-3/2} + 3x^2\epsilon^{-1}(y_0 + \epsilon^{1/2} y_1) + 3y_0^2(x\epsilon^{1/2})]$$
$$+ x\epsilon^{-1/2} + y_0 + \epsilon^{1/2} y_1 - 2 = 0$$

Collecting terms with like powers of ϵ, we have

$$\epsilon^{-1/2}(x^3 + x) + \epsilon^0(3x^2y_0 + y_0 - 2) + \epsilon^{1/2}(3x^2y_1 + 3xy_0^2 + y_1) = 0$$

The resulting sequence of algebraic equations is as follows

$$\epsilon^{-1/2}: \quad x^3 + x = 0 \tag{3.9}$$

$$\epsilon^0: \quad 3x^2y_0 + y_0 + y_0 - 2 = 0 \tag{3.10}$$

$$\epsilon^{1/2}: \quad 3x^2y_1 + 3xy_0^2 + y_1 = 0 \tag{3.11}$$

The solutions of Eq. (3.9) are

$$x = 0 \qquad x = \pm i$$

Discarding the solution $x = 0$, which corresponds to the real root, we use $x = \pm i$ in Eqs. (3.10) and (3.11) to solve for y_0 and y_1 and obtain

$$y_0 = -1 \qquad y_1 = \pm \tfrac{3}{2} i$$

Thus, the complex roots are given by

$$y = i\epsilon^{-1/2} - 1 + \tfrac{3}{2} i\epsilon^{1/2} \tag{3.12}$$

$$y = -i\epsilon^{-1/2} - 1 - \tfrac{3}{2} i\epsilon^{1/2} \tag{3.13}$$

3.3 A TRANSCENDENTAL EQUATION

For transient conduction in a slab with convective boundary conditions, it is required to find the roots of the transcendental equation

$$z \tan z = \mathrm{Bi} \tag{3.14}$$

where $\mathrm{Bi} = hL/k$ is the Biot number. For small Bi, let us assume an expansion of the form

$$z = z_0 + \mathrm{Bi}\, z_1 + \mathrm{Bi}^2 z_2 + \mathrm{Bi}^3 z_3 + O(\mathrm{Bi}^4) \tag{3.15}$$

Substituting Eq. (3.15) into Eq. (3.14) and proceeding as in Section 2.3, it is easy to show that

$$z = z_0 + \mathrm{Bi}\,\frac{1}{z_0} - \mathrm{Bi}^2\,\frac{1}{z_0^3} + \mathrm{Bi}^3\left(\frac{2}{z_0^5} - \frac{1}{3z_0^3}\right) + O(\mathrm{Bi}^4) \tag{3.16}$$

where $\tan z_0 = 0$ or $z_0 = 0, \pi, \ldots, n\pi$, n being a positive integer. The expansion in Eq. (3.16) is valid for all roots except the first one corresponding to $z_0 = 0$. For $z_0 = 0$, the second term in Eq. (3.16) becomes singular. As we go to higher-order terms, the singularity grows progressively stronger. The occurrence of growing singularity is a defect which is encountered in many problems. One possible remedy is to use the method of strained coordinates which is the subject of Chapter 4.

3.4 A FIRST–ORDER DIFFERENTIAL EQUATION

Consider a simple first-order, nonlinear differential equation with a specified condition

$$(x + \epsilon y)\frac{dy}{dx} + y = 0 \tag{3.17}$$

$$x = 1 \qquad y = 1 \tag{3.18}$$

For small ϵ, let us carry out a perturbation expansion of the form

$$y = y_0 + \epsilon y_1 + \epsilon^2 y_2 \tag{3.19}$$

Substituting Eq. (3.19) into Eqs. (3.17) and (3.18) and equating coefficients of like powers of ϵ, we have

$$\epsilon^0: \quad x\frac{dy_0}{dx} + y_0 = 0 \tag{3.20}$$

$$x = 1 \qquad y_0 = 1 \tag{3.21}$$

$$\epsilon^1: \quad x\frac{dy_1}{dx} + y_1 = -y_0\frac{dy_0}{dx} \tag{3.22}$$

$$x = 1 \qquad y_1 = 0 \tag{3.23}$$

$$\epsilon^2: \quad x\frac{dy_2}{dx} + y_2 = -y_0\frac{dy_1}{dx} - y_1\frac{dy_0}{dx} \tag{3.24}$$

$$x = 1 \qquad y_2 = 0 \tag{3.25}$$

The solutions of Eqs. (3.20)–(3.25) are readily obtained as

$$y_0 = \frac{1}{x} \qquad y_1 = \frac{x^2 - 1}{2x^3} \qquad y_2 = -\frac{x^2 - 1}{2x^5} \tag{3.26}$$

Hence, the expansion in Eq. (3.19) becomes

$$y = \frac{1}{x} + \epsilon\frac{x^2 - 1}{2x^3} - \epsilon^2\frac{x^2 - 1}{2x^5} \tag{3.27}$$

The defect of the solution is that, at $x = 0, y_0$ is singular, and the singularity grows stronger in y_1 and y_2 because of the higher powers of x in the denominator. As it stands, the solution is computationally useless in the neighborhood of $x = 0$.

To gain insight into the behavior of y as $x \to 0$, we derive the exact solution of Eqs. (3.17) and (3.18). Dividing Eq. (3.17) by dy/dx we have

$$y \frac{dx}{dy} + x = -\epsilon y$$

which can be written as

$$\frac{d}{dy}(xy) = -\epsilon y \tag{3.28}$$

Integrating Eq. (3.28) and imposing the condition in Eq. (3.18) gives

$$xy = 1 + \tfrac{1}{2}\epsilon(1 - y^2)$$

which may be rearranged to give

$$y^2 + \frac{2}{\epsilon} xy = \frac{2}{\epsilon} + 1$$

or

$$\left(y + \frac{x}{\epsilon}\right)^2 = \frac{x^2}{\epsilon^2} + \frac{2}{\epsilon} + 1$$

or

$$y = -\frac{x}{\epsilon} + \left(\frac{x^2}{\epsilon^2} + \frac{2}{\epsilon} + 1\right)^{1/2} \tag{3.29}$$

From Eq. (3.29) it follows that $y \to [(2/\epsilon) + 1]^{1/2}$ as $x \to 0$. In Fig. 3.1 we plot the one-term, two-term, and three-term perturbation solutions and compare them with the exact solution. For $x < 0.6$, the perturbation solution begins to deviate from the exact solution, and this deviation is disastrous as the limit $x = 0$ is approached.

To trace the source of nonuniformity, let us expand Eq. (3.29) binomially and then rearrange the result to get

$$y = 1 + \epsilon^{-1}(1 - x) + \epsilon^{-2}\tfrac{1}{2}(x^2 - 1) + O(\epsilon^{-3}) \tag{3.30}$$

which shows that the correct form of expansion is a series in inverse powers of ϵ and not direct powers as stipulated in Eq. (3.19).

Instead of Eq. (3.19), we now assume an expansion of the form

$$y = y_0 + \epsilon^{-1} y_1 + \epsilon^{-2} y_2 \tag{3.31}$$

and substitute this into Eqs. (3.17) and (3.18). Equating coefficients of ϵ, ϵ^0, and ϵ^{-1} to zero, we get

$$\epsilon: \quad y_0 \frac{dy_0}{dx} = 0 \tag{3.32}$$

$$x = 1 \quad y_0 = 1 \tag{3.33}$$

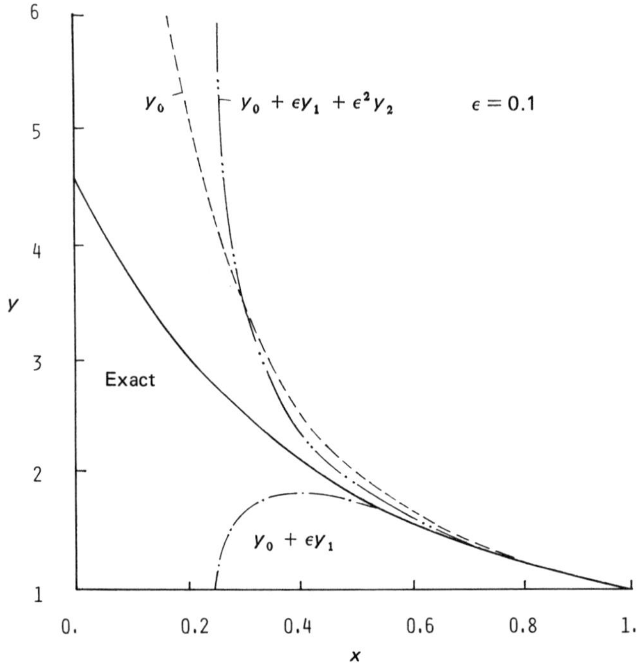

Figure 3.1 Perturbation and exact solutions of Eqs. (3.17) and (3.18).

$$\epsilon^0: \quad y_0 \frac{dy_1}{dx} + \frac{dy_0}{dx} y_1 = -x \frac{dy_0}{dx} - y_0 \tag{3.34}$$

$$x = 1 \qquad y_1 = 0 \tag{3.35}$$

$$\epsilon^{-1}: \quad y_0 \frac{dy_2}{dx} + \frac{dy_0}{dx} y_2 = -x \frac{dy_1}{dx} - y_1 \frac{dy_1}{dx} - y_1 \tag{3.36}$$

$$x = 1 \qquad y_2 = 0 \tag{3.37}$$

The solutions of Eqs. (3.32)–(3.37) are readily obtained as

$$y_0 = 1 \qquad y_1 = 1 - x \qquad y_2 = \tfrac{1}{2}(x^2 - 1)$$

Substituting in Eq. (3.31) it follows that

$$y = 1 + \epsilon^{-1}(1 - x) + \epsilon^{-2}\tfrac{1}{2}(x^2 - 1) \tag{3.38}$$

which is identical to the expanded version of the exact solution in Eq. (3.30).

3.5 RADIATING HEAT SHIELD

A common device used to protect spacecraft during re-entry is a heat shield. It takes the form of a shell made of conducting material attached to the exposed surface of the spacecraft. For a typical heat shield employed on the wing (see Fig. 3.2), a simple energy balance gives

$$k\delta \frac{d^2 t}{dx^2} - \sigma E t^4 + Q(x) = 0$$

where t = temperature of the solid
k = heat conductivity
δ = thickness of the shield
σ = Stefan-Boltzmann constant
E = surface emissivity

The function $Q(x)$ represents the aerodynamic heat transfer distribution normal with respect to radiation.

By introducing the dimensionless variables $\theta = t/\bar{t}$, $X = x/L$, $q(X) = Q(x)/(\sigma E \bar{t}^4)$, and $\epsilon = k\delta/\sigma E L^2 \bar{t}^3$, the following boundary value problem is obtained:

$$\epsilon \frac{d^2\theta}{dX^2} - \theta^4 + q(X) = 0 \tag{3.39}$$

$$X = 0 \quad \frac{d\theta}{dX} = 0$$
$$X = 1 \quad \frac{d\theta}{dX} = 0 \tag{3.40}$$

As usual, let us try an expansion of the form

$$\theta = \theta_0 + \epsilon\theta_1 \tag{3.41}$$

which when introduced into Eqs. (3.39) and (3.40) leads to the following sequence of problems

$$\epsilon^0: \quad \theta_0^4 = q(X) \tag{3.42}$$

$$X = 0 \quad \frac{d\theta_0}{dX} = 0$$
$$X = 1 \quad \frac{d\theta_0}{dX} = 0 \tag{3.43}$$

$$\epsilon^1: \quad 4\theta_0^3\theta_1 = \frac{d^2\theta_0}{dX^2} \tag{3.44}$$

$$X = 0 \quad \frac{d\theta_1}{dX} = 0$$
$$X = 1 \quad \frac{d\theta_1}{dX} = 0 \tag{3.45}$$

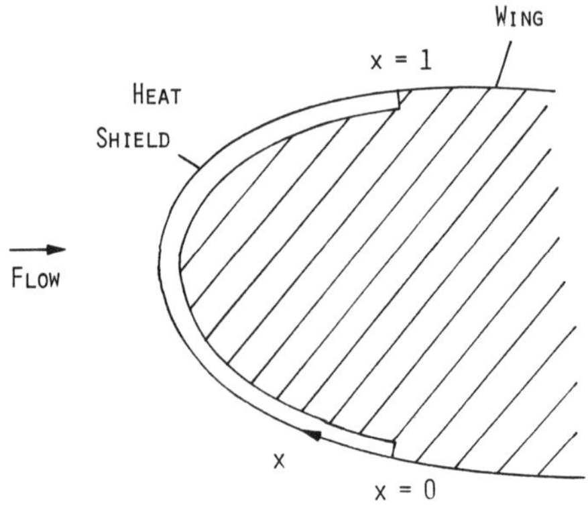

Figure 3.2 Radiating heat shield.

Solving for θ_0 and θ_1, and using them in Eq. (3.41) we have

$$\theta = [q(X)]^{1/4} + \frac{1}{16}\,\epsilon\,[q(X)]^{-3/2}\left\{\frac{d^2q(X)}{dX^2} - \frac{3}{4}\,[q(X)]^{-1}\,\frac{dq(X)}{dX}\right\} \tag{3.46}$$

The expansion in Eq. (3.46) cannot satisfy the boundary conditions in Eq. (3.40) and consequently fails to describe the temperature distribution near the ends, i.e., $X = 0$ and $X = 1$. The reason for the failure is the appearance of ϵ in front of the highest derivative in Eq. (3.39). In the process of expansion, the original second-order differential equation was simply reduced to two algebraic equations, Eqs. (3.42) and (3.44). We thus lost the second derivative and the associated boundary conditions. The resolution to this class of problems where ϵ is a coefficient of the highest derivative is best achieved with the method of matched asymptotic expansions discussed in Chapter 5. The solution to the present problem is offered as an exercise at the end of Chapter 5.

3.6 MELTING OF A FINITE SLAB

Goodman and Shea (1960) have studied the melting rate of a finite slab due to constant heat flux at one face, the other face being insulated. Using the heat balance integral method, the average temperature in the melted region, θ_1, and the solid region, θ_2, and the thickness of the melted region, $s(t)$, are found to be governed by the following equations:

$$\rho_2 L\,\frac{ds}{dt} + \frac{k_1}{\kappa_1}\,\frac{d\theta_1}{dt} + \frac{k_2}{\kappa_2}\,\frac{d\theta_2}{dt} = H$$

$$\theta_1 = s^2\left(\frac{H}{2k_1} - \frac{1}{3\kappa_1}\,\frac{d\theta_1}{dt}\right)$$

$$\theta_2 = -\frac{(l-s)^2}{3\kappa_2}\,\frac{d\theta_2}{dt}$$

with the initial conditions

$$s(t_m) = 0 \qquad \theta_1(t_m) = 0 \qquad \theta_2(t_m) = -\frac{Hl^2}{3k_2}$$

where ρ = density
L = latent heat of melting
k = thermal conductivity
κ = diffusivity
H = heat flux at $x = 0$

The equations are derived by space-averaging of the energy equations of the two layers, including appropriate energy balance along the melting line. Let us next introduce the following dimensionless quantities

$$u = \frac{s}{l} \qquad w = \frac{k_1 \kappa_2 \theta_1}{k_2 \kappa_1 Vl} \qquad v = \frac{\theta_2}{Vl} \qquad \tau = \frac{\kappa_2}{l^2}(t - t_m)$$

and the following dimensionless parameters

$$\epsilon = \frac{Hl}{2k_2 V} \qquad \mu = \frac{k_2 V}{\kappa_2 \rho_2 L} \qquad \nu = \frac{\kappa_1}{\kappa_2}$$

where l, $-V$, and t_m are the thickness of the two regions, the initial temperature of the solid, and the time at which melting begins, respectively. The problem is then reduced to the solution of the following boundary value problem

$$\frac{du}{d\tau} + \mu\frac{dv}{d\tau} + \frac{dw}{d\tau} = 2\mu\epsilon \tag{3.47}$$

$$(1 - u)^2 \frac{dv}{d\tau} = -3v \tag{3.48}$$

$$\frac{1}{3}u^2 \frac{dw}{d\tau} = \epsilon u^3 - \nu w \tag{3.49}$$

$$\tau = 0 \qquad u = 0 \qquad v = \tfrac{2}{3}\epsilon \qquad w = 0 \tag{3.50}$$

where u = the location of melting-front
v = temperature distribution in the liquid
w = temperature distribution in the solid
μ = Stefan number
τ = time
ϵ = heat flux

For small ϵ, assume

$$u = u_0 + \epsilon u_1 + \epsilon^2 u_2 \tag{3.51}$$

$$v = v_0 + \epsilon v_1 + \epsilon^2 v_2 \tag{3.52}$$

$$w = w_0 + \epsilon w_1 + \epsilon^2 w_2 \tag{3.53}$$

Following the usual procedure, we obtain the following system of equations

$$\epsilon^0: \quad \frac{du_0}{d\tau} + \mu \frac{dv_0}{d\tau} + \mu \frac{dw_0}{d\tau} = 0 \tag{3.54}$$

$$(1 - u_0)^2 \frac{dv_0}{d\tau} = -3v_0 \tag{3.55}$$

$$\frac{1}{3} u_0^2 \frac{dw_0}{d\tau} = -v_0 w_0 \tag{3.56}$$

$$\tau = 0 \quad u_0 = 0 \quad v_0 = 0 \quad w_0 = 0 \tag{3.57}$$

$$\epsilon^1: \quad \frac{du_1}{d\tau} + \mu \frac{dv_1}{d\tau} + \mu \frac{dw_1}{d\tau} = 2\mu \tag{3.58}$$

$$(1 - u_0)^2 \frac{dv_1}{d\tau} - 2u_1(1 - u_0) \frac{dv_0}{d\tau} = -3v_1 \tag{3.59}$$

$$\frac{1}{3} u_0^2 \frac{dw_1}{d\tau} + \frac{2}{3} u_0 u_1 \frac{dw_0}{d\tau} = u_0^2 - (v_0 w_1 + w_0 v_1) \tag{3.60}$$

$$\tau = 0 \quad u_1 = 0 \quad v_1 = -\tfrac{2}{3} \quad w_1 = 0 \tag{3.61}$$

$$\epsilon^2: \quad \frac{du_2}{d\tau} + \mu \frac{dv_2}{d\tau} + \mu \frac{dw_2}{d\tau} = 0 \tag{3.62}$$

$$(1 - u_0)^2 \frac{dv_2}{d\tau} - 2u_1(1 - u_0) \frac{dv_1}{d\tau} + (u_1^2 + 2u_0 u_2 - 2u_2) \frac{dv_0}{d\tau} = -3v_2 \tag{3.63}$$

$$\frac{1}{3} u_0^2 \frac{dw_2}{d\tau} + \frac{2}{3} u_0 u_1 \frac{dw_1}{d\tau} + \frac{1}{3} (2u_0 u_2 + u_1^2) \frac{dw_0}{d\tau}$$
$$= 2u_0 u_1 - (v_0 w_2 + v_1 w_1 + w_0 v_2) \tag{3.64}$$

$$\tau = 0 \quad u_2 = 0 \quad v_2 = 0 \quad w_2 = 0 \tag{3.65}$$

Zero-Order Solution

First, examine Eq. (3.55). Since $v_0(0) = 0$, it follows that $(dv_0/d\tau)(0) = 0$ and v_0 is identically zero. Next, integrating Eq. (3.54) subject to Eq. (3.57) gives $u_0 + \mu v_0 + \mu w_0 = 0$, which in view of $v_0 = 0$, becomes $u_0 + \mu w_0 = 0$ or $u_0 = -\mu w_0$. Substituting for u_0 in Eq. (3.56) gives $\frac{1}{6}\mu^2 (dw_0^3/d\tau) = -v_0 = 0$, from which it follows that $w_0 = 0$, and hence $u_0 = 0$. Thus,

$$u_0 = v_0 = w_0 = 0 \tag{3.66}$$

First-Order Solution

In light of Eq. (3.66), Eq. (3.59) reduces to

$$\frac{dv_1}{d\tau} = -3v_1 \tag{3.67}$$

while the solution for w_1 from Eq. (3.60) follows as

$$w_1 = 0 \tag{3.68}$$

which in turn reduces Eq. (3.58) to

$$\frac{du_1}{d\tau} + \mu \frac{dv_1}{d\tau} = 2\mu \tag{3.69}$$

The solution of Eq. (3.67) subject to the initial conditions given in Eq. (3.61) is

$$v_1 = -\tfrac{2}{3} e^{-3\tau} \tag{3.70}$$

Using Eq. (3.70) in Eq. (3.69) and integrating, the solution for u_1 satisfying Eq. (3.61) can be obtained as

$$u_1 = 2\mu\tau + \tfrac{2}{3}\mu(e^{-3\tau} - 1) \tag{3.71}$$

Second-Order Solution

Employing the zero-order and first-order solutions, it follows readily from Eq. (3.64) that

$$w_2 = 0 \tag{3.72}$$

while Eq. (3.63) reduces to

$$\frac{dv_2}{d\tau} + 3v_2 = \frac{8}{3}\mu e^{-3\tau}(e^{-3\tau} - 1) + 8\mu\tau e^{-3\tau} \tag{3.73}$$

The solution of Eq. (3.73) subject to Eq. (3.65) is

$$v_2 = \tfrac{8}{9}\mu(e^{-3\tau} - e^{-6\tau} - 3\tau e^{-3\tau} + \tfrac{9}{2}\tau^2 e^{-3\tau}) \tag{3.74}$$

The solution of u_2 from Eq. (3.62) is

$$u_2 = -\mu v_2 = -\tfrac{8}{9}\mu^2(e^{-3\tau} - e^{-6\tau} - 3\tau e^{-3\tau} + \tfrac{9}{2}\tau^2 e^{-3\tau}) \tag{3.75}$$

Thus the complete solution up to $O(\epsilon^2)$ is

$$u = \epsilon[2\mu\tau + \tfrac{2}{3}\mu(e^{-3\tau} - 1)] - \epsilon^2 \tfrac{8}{9}\mu^2(e^{-3\tau} - e^{-6\tau} - 3\tau e^{-3\tau} + \tfrac{9}{2}\tau^2 e^{-3\tau}) + O(\epsilon^3) \tag{3.76}$$

$$v = -\epsilon \tfrac{2}{3} e^{-3\tau} + \epsilon^2 \tfrac{8}{9} \mu^2 (e^{-3\tau} - e^{-6\tau} - 3\tau e^{-3\tau} + \tfrac{9}{2}\tau^2 e^{-3\tau}) + O(\epsilon^3) \qquad (3.77)$$

$$w = 0 + O(\epsilon^3) \qquad (3.78)$$

It is instructive to examine the foregoing solutions in light of the physics of the problem. Since $\epsilon = 0$ means zero heat flux input, the zero-order solutions in Eq. (3.66), implying no temperature rise in the solid, are logical. Likewise, the first-order solutions, Eqs. (3.68), (3.70), and (3.71), are well behaved and pose no difficulty. However, the solution for v_2 contains terms like $\tau e^{-3\tau}$ and $\tau^2 e^{-3\tau}$ which for large τ make v positive and thus violate the physics of the problem which demands that v be negative. The terms $\tau e^{-3\tau}$ and $\tau^2 e^{-3\tau}$ resemble the "secular terms" arising in nonlinear oscillations. In Chapter 4 we use the method of strained coordinates to remove these undesirable terms.

3.7 INWARD SPHERICAL SOLIDIFICATION

As a last example, we consider the freezing of a saturated liquid in a spherical container due to lowering of its wall temperature. This problem has been considered, among others, by Pedroso and Domoto (1973a) who write the transient heat conduction equation in spherical form as

$$\frac{\partial T}{\partial t} = \frac{\alpha}{R} \frac{\partial^2 (RT)}{\partial R^2}$$

The constant temperatures at the fixed wall ($R = R_w$), T_w, and at the freezing front ($R = R_f$), T_f, yield the boundary conditions

$$R = R_w \qquad T = T_w$$

$$R = R_f \qquad T = T_f$$

The energy balance at the interface leads to

$$\frac{dR_f}{dt} = \frac{k}{\rho L} \left(\frac{\partial T}{\partial R}\right)_{R=R_f}$$

By introducing the dimensionless quantities

$$\epsilon = C \frac{T_f - T_w}{L} \qquad u = \frac{T - T_w}{T_f - T_w} \qquad r = \frac{R}{R_w}$$

$$r_f = \frac{R_f}{R_w} \qquad \tau = k \frac{T_f - T_w}{\rho L R_w^2} t$$

the boundary value problem for this problem becomes

$$\frac{1}{r}\frac{\partial^2(ur)}{\partial r^2} = \epsilon \frac{\partial u}{\partial r_f}\frac{\partial u}{\partial r}\bigg|_{r=r_f} \tag{3.79}$$

$$u(r_f, r=1) = 0 \qquad u(r_f, r=r_f) = 1 \tag{3.80}$$

$$\frac{dr_f}{d\tau} = \frac{\partial u}{\partial r}\bigg|_{r=r_f} \tag{3.81}$$

We can work out a three-term perturbation solution by assuming

$$u = u_0 + \epsilon u_2 + \epsilon^2 u_2 \tag{3.82}$$

Substituting Eq. (3.82) into Eqs. (3.79)-(3.81), the following system of equations for u_0, u_1, and u_2 is obtained

$$\epsilon^0: \quad \frac{1}{r}\frac{\partial^2(u_0 r)}{\partial r^2} = 0 \tag{3.83}$$

$$u_0(r_f, r=1) = 0 \qquad u_0(r_f, r=r_f) = 1 \tag{3.84}$$

$$\epsilon^1: \quad \frac{1}{r}\frac{\partial^2(u_1 r)}{\partial r^2} = \frac{\partial u_0}{\partial r_f}\frac{\partial u_0}{\partial r}\bigg|_{r=r_f} \tag{3.85}$$

$$u_1(r_f, r=1) = 0 \qquad u_1(r_f, r=r_f) = 0 \tag{3.86}$$

$$\epsilon^2: \quad \frac{1}{r}\frac{\partial^2(u_2 r)}{\partial r^2} = \frac{\partial u_0}{\partial r_f}\frac{\partial u_1}{\partial r}\bigg|_{r=r_f} + \frac{\partial u_0}{\partial r}\bigg|_{r=r_f}\frac{\partial u_1}{\partial r_f} \tag{3.87}$$

$$u_2(r_f, r=1) = 0 \qquad u_2(r_f, r=r_f) = 0 \tag{3.88}$$

Zero-order solution. Integrating Eq. (3.83) twice, we have

$$u_0 = f_1 + \frac{f_2}{r}$$

where f_1 and f_2 are functions of r_f. Applying the boundary conditions in Eq. (3.84) gives

$$f_1 = \frac{1}{1-(1/r_f)} \qquad f_2 = -f_1$$

Thus

$$u_0 = \frac{1-1/r}{1-1/r_f} \tag{3.89}$$

First-order solution. From Eq. (3.89), we get

$$\frac{\partial u_0}{\partial r_f} = -\frac{1-1/r}{(r_f-1)^2}$$

$$\frac{\partial u_0}{\partial r} = \frac{1/r^2}{1 - 1/r_f} \qquad \left.\frac{\partial u_0}{\partial r}\right|_{r=r_f} = \frac{1}{r_f(r_f - 1)}$$

The first-order problem in Eq. (3.85) now becomes

$$\frac{\partial^2 (u_1 r)}{\partial r^2} = -\frac{r - 1}{r_f (r_f - 1)^3} \tag{3.90}$$

Integrating Eq. (3.90) twice

$$u_1 r = -\frac{1}{r_f (r_f - 1)^3} \left(\frac{1}{6} r^3 - \frac{1}{2} r^2 \right) + f_3 r + f_4$$

where f_3 and f_4 are functions of r_f. We again apply the boundary conditions, Eq. (3.86), and obtain

$$f_3 = \frac{r_f^3 - 3r_f^2 + 2}{6r_f (r_f - 1)^4} \qquad f_4 = -\frac{r_f^2 - 3r_f + 2}{6(r_f - 1)^4}$$

The solution for u_1 becomes

$$u_1 = \frac{r_f^2 - 3r_f + 2}{6(r_f - 1)^4} \left(1 - \frac{1}{r} \right) - \frac{r^2 - 3r + 2}{6r_f (r_f - 1)^3} \tag{3.91}$$

Second-order solution. From Eq. (3.91) we have

$$\frac{\partial u_1}{\partial r_f} = \frac{-2r_f + 5}{6(r_f - 1)^4} \left(1 - \frac{1}{r} \right) + \frac{4r_f - 1}{6r_f^2 (r_f - 1)^4} (r^2 - 3r + 2)$$

$$\frac{\partial u_1}{\partial r} = \frac{r_f^2 - 3r_f + 2}{6(r_f - 1)^4} \frac{1}{r^2} - \frac{2r - 3}{6r_f (r_f - 1)^3}$$

$$\left.\frac{\partial u_1}{\partial r}\right|_{r=r_f} = \frac{1 - r_f^3 + 3r_f^2 - 3r_f}{3r_f^2 (r_f - 1)^4} = -\frac{1}{3r_f^2 (r_f - 1)}$$

The second-order problem simplifies to

$$\frac{1}{r} \frac{\partial^2 (u_2 r)}{\partial r^2} = \frac{r_f + 2}{6r_f^2 (r_f - 1)^5} \left(1 - \frac{1}{r} \right) + \frac{4r_f - 1}{6r_f^3 (r_f - 1)^5} (r^2 - 3r + 2) \tag{3.92}$$

Integrating Eq. (3.92) twice and imposing the boundary conditions in Eq. (3.88), the solution for u_2 becomes

$$u_2 = \frac{-12r_f^5 + 53r_f^4 - 85r_f^3 + 60r_f^2 - 8r_f - 8}{360r_f^3 (r_f - 1)^6} \left(1 - \frac{1}{r} \right)$$
$$+ \frac{r_f + 2}{36r_f^2 (r_f - 1)^5} (r^2 - 3r + 2) + \frac{4r_f - 1}{360r_f^3 (r_f - 1)^5} (3r^4 - 15r^3 + 20r^2 - 8) \tag{3.93}$$

On examining Eq. (3.91) it is seen that as $r_f \to 0$, that is, as the freezing front approaches the center, the second term of Eq. (3.91) makes u_1 singular. This singularity grows stronger in the solution for u_2 as higher powers of r_f appear in the denominators of Eq. (3.93).

We now derive the freezing time solution. Using Eq. (3.82) in Eq. (3.81) we have

$$\frac{dr_f}{d\tau} = \left.\frac{\partial u_0}{\partial r}\right|_{r=r_f} + \epsilon \left.\frac{\partial u_1}{\partial r}\right|_{r=r_f} + \epsilon^2 \left.\frac{\partial u_2}{\partial r}\right|_{r=r_f} \tag{3.94}$$

From the zero-order and the first-order solutions, $(\partial u_0/\partial r)|_{r=r_f}$ and $(\partial u_1/\partial r)|_{r=r_f}$ are known. To obtain $(\partial u_2/\partial r)|_{r=r_f}$, we differentiate Eq. (3.93) and evaluate it at $r = r_f$ to get

$$\left.\frac{\partial u_2}{\partial r}\right|_{r=r_f} = \frac{6r_f^6 - 29r_f^5 + 55r_f^4 - 50r_f^3 + 20r_f^2 - r_f - 1}{45r_f^4(r_f - 1)^6}$$

in which the numerator simplifies to $(1 + 6r_f)(r_f - 1)^5$ giving

$$\left.\frac{\partial u_2}{\partial r}\right|_{r=r_f} = \frac{1 + 6r_f}{45r_f^4(r_f - 1)} \tag{3.95}$$

Using the appropriate expressions for $(\partial u_0/\partial r)|_{r=r_f}$, $(\partial u_1/\partial r)|_{r=r_f}$, and $(\partial u_2/\partial r)|_{r=r_f}$, Eq. (3.94) takes the form

$$\frac{dr_f}{d\tau} = \frac{1}{r_f(r_f - 1)}\left(1 - \epsilon \frac{1}{3r_f} + \epsilon^2 \frac{1 + 6r_f}{45r_f^3}\right) \tag{3.96}$$

Separating the variables in Eq. (3.96) gives

$$r_f(r_f - 1)\left(1 - \epsilon \frac{1}{3r_f} + \epsilon^2 \frac{1 + 6r_f}{45_f^3}\right)^{-1} dr_f = d\tau$$

Expanding the term in parentheses binomially, and retaining terms up to $O(\epsilon^2)$, we have

$$r_f(r_f - 1) + \epsilon \frac{1}{3}(r_f - 1) - \epsilon^2 \frac{1}{45}\left(1 - \frac{1}{r_f^2}\right) = d\tau \tag{3.97}$$

Integrating Eq. (3.97) and using the initial condition

$$r_f = 1 \qquad \tau = 0 \tag{3.98}$$

gives the solution for τ as

$$\tau = \frac{2r_f^3 - 3r_f^2 + 1}{6} + \epsilon \frac{1}{6}(r_f - 1)^2 - \epsilon^2 \frac{1}{45}\frac{(r_f - 1)^2}{r_f} \tag{3.99}$$

The solution in Eq. (3.99) is plotted in Fig. 3.3 for $\epsilon = 0.1$, 0.5, and 1.0. In each case the solution diverges as r_f approaches zero. Thus, we cannot use Eq.

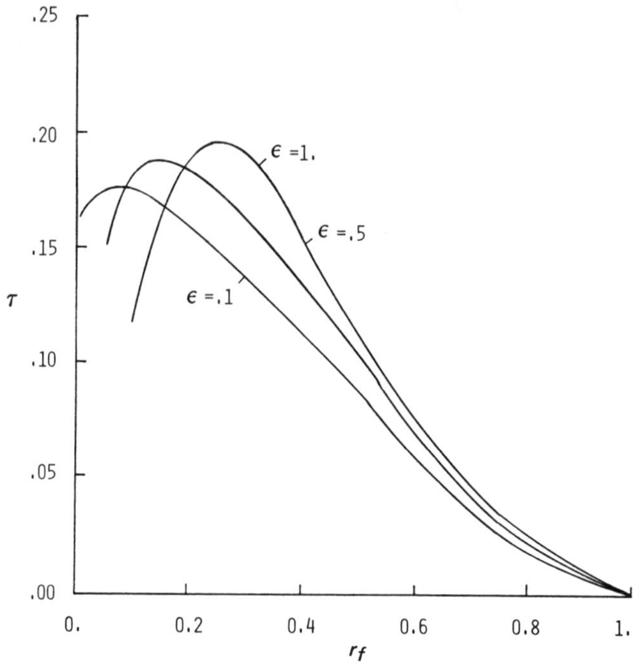

Figure 3.3 Freezing time in inward spherical solidification.

(3.99) to predict the time for complete solidification ($r_f = 0$). However, the singularity in Eq. (3.99) first appears in the third term; the two-term solution for τ is therefore uniformly valid giving

$$\tau = \tfrac{1}{6}(1 + \epsilon) \quad \text{at } r_f = 0 \tag{3.100}$$

The conclusion from the above analysis is that except for the zero-order term for u series and zero- and first-order terms for τ series, all higher-order terms are singular as $r_f \to 0$. In Chapter 4 we pick up this problem again and use the method of strained coordinates to derive a uniformly valid solution.

3.8 COOLING OF A LUMPED SYSTEM WITH VARIABLE HEAT TRANSFER COEFFICIENT

In this section we will attempt solutions of Eqs. (1.61) and (1.62). As noted there, ϵ is small and therefore we seek a solution of the form

$$\theta = \theta_0 + \epsilon\theta_1 + \epsilon^2\theta_2 \tag{3.101}$$

If Eq. (3.101) is substituted into Eq. (1.61), the term θ^ϵ gives rise to $(\theta_0 + \epsilon\theta_1 + \epsilon^2\theta_2)^\epsilon$ which obviously is not fully explicit in ϵ. To overcome this difficulty, the term is written as exp $[\epsilon \ln (\theta_0 + \epsilon\theta_1 + \epsilon^2\theta_2)]$ and then expanded. Thus

$$(\theta_0 + \epsilon\theta_1 + \epsilon^2\theta_2)^\epsilon = \exp [\epsilon \ln (\theta_0 + \epsilon\theta_1 + \epsilon^2\theta_2)]$$

$$= 1 + \epsilon \ln \theta_0 + \epsilon^2 \left[\frac{\theta_1}{\theta_0} + \frac{1}{2} (\ln \theta_0)^2 \right] + O(\epsilon^3) \quad (3.102)$$

Although we have succeeded in forcing a small fractional power into a series in powers of the exponent, the expansion shown in Eq. (3.102) is not uniformly valid. This nonuniformity will exhibit itself in the final solution.

With the use of Eq. (3.102) we can now write the sequence of perturbation equations as

$$\epsilon^0: \quad \frac{d\theta_0}{d\tau} + \theta_0 = 0 \quad (3.103)$$

$$\tau = 0 \qquad \theta_0 = 1 \quad (3.104)$$

$$\epsilon^1: \quad \frac{d\theta_1}{d\tau} + \theta_1 + \theta_0 \ln \theta_0 = 0 \quad (3.105)$$

$$\tau = 0 \qquad \theta_1 = 0 \quad (3.106)$$

$$\epsilon^2: \quad \frac{d\theta_2}{d\tau} + \theta_2 + \theta_1(1 + \ln \theta_0) + \frac{1}{2} \theta_0(\ln \theta_0)^2 \quad (3.107)$$

$$\tau = 0 \qquad \theta_2 = 0 \quad (3.108)$$

Solving for θ_0, θ_1, and θ_2 the solution for θ becomes

$$\theta = e^{-\tau} + \tfrac{1}{2}\epsilon\tau^2 e^{-\tau} + \tfrac{1}{3}\epsilon^2(\tfrac{3}{8}\tau^4 - \tau^3)e^{-\tau} + O(\epsilon^3) \quad (3.109)$$

The exact solution of Eqs. (1.61) and (1.62) is easily obtained as

$$\theta = (1 + \epsilon\tau)^{-1/\epsilon} \qquad \epsilon \neq 0 \quad (3.110)$$

Examining Eq. (3.109) it is seen that if $\tau = O(\sqrt{2/\epsilon})$, the second term is of the same order as the first term which violates the basic assumption that $\epsilon\theta_1$ is small compared to θ_0. The agreement between the perturbation solution in Eq. (3.109) and the exact solution in Eq. (3.110) is good as long as τ does not exceed $\sqrt{2/\epsilon}$.

PROBLEMS

3.1 Determine three terms in the expansion of each of the roots of

$$\epsilon y^3 + (y - 1)^2 = 0$$

3.2 Verify that a two-term expansion for

$$(x+\epsilon y)\frac{dy}{dx}+y=2x$$

$$x=1 \qquad y=b>1$$

leads to

$$y=x+\frac{b-1}{x}+\epsilon\,\frac{1}{2x}\left[(b-1)^2-\frac{(b-1)^2}{x^2}-x^2\right]+O(\epsilon^2)$$

3.3 Carry out a regular perturbation expansion for

$$(x+\epsilon y)\frac{dy}{dx}-\frac{1}{2}y=1+x^2$$

$$x=1 \qquad y=1$$

and show that

$$y=\tfrac{1}{3}(2x^2+7x^{1/2}-6)-\epsilon\tfrac{1}{45}(16x^3-175x^{3/2}-240x+539x^{1/2}$$
$$+105x^{-1/2}-245)+O(\epsilon^2)$$

Identify the singular term and its region of nonuniformity.

3.4 Derive a two-term expansion of

$$(x+\epsilon y)\frac{dy}{dx}+(2+x)y=0$$

$$x=1 \qquad y=\frac{1}{e}$$

as
$$y=\frac{e^{-x}}{x^2}+\frac{e^{-x}}{x^2}\int_1^x e^{-u}\left(\frac{2}{u^4}+\frac{1}{u^3}\right)du+O(\epsilon^2)$$

Discuss its uniformity in the region of $x=0$.

3.5 The mathematical model for freezing in cylindrical geometry with constant wall temperature is (Asfar et al., 1979)

$$\frac{1}{r}\frac{\partial}{\partial r}\left(r\frac{\partial u}{\partial r}\right)=\epsilon\left.\frac{\partial u}{\partial r_f}\frac{\partial u}{\partial r}\right|_{r=r_f}$$

$$u(r_f, r=1)=0 \qquad u(r_f, r=r_f)=1$$

$$\frac{dr_f}{d\tau}=\left.\frac{\partial u}{\partial r}\right|_{r=r_f}$$

Develop a regular perturbation solution and demonstrate that

$$u = \frac{\ln r}{\ln r_f} + \epsilon \left[\frac{1 + r_f^2 (\ln r_f - 1)}{4 r_f^2 (\ln r_f)^4} \ln r - \frac{1 + r^2 (\ln r - 1)}{4 r_f^2 (\ln r_f)^3} \right] + O(\epsilon^2)$$

$$\tau = \frac{1}{2} r_f^2 \ln r_f + \frac{1}{4} (1 - r_f^2) + \epsilon \frac{1}{4} (1 - r_f^2) \left(1 + \frac{1}{\ln r_f} \right) + O(\epsilon^2)$$

Discuss the validity of the above solution for outward and inward freezing.

3.6 Obtain a two-term regular perturbation solution of Eq. (3.79) subject to

$$u(r_f, r = 1) = -\frac{1}{\text{Bi}} \frac{\partial u}{\partial r}\bigg|_{r=1} \qquad u(r_f, r = r_f) = 1$$

which is applicable to convective cooling of the spherical container. The solution should read as

$$u = u_0 + \epsilon u_1$$

where

$$u_0 = \frac{1 - \text{Bi}^* (1/r)}{1 - \text{Bi}^* (1/r_f)}$$

and

$$u_1 = \left[\frac{\text{Bi}^{*3} (r_f^2 - 3\,\text{Bi}^* r_f) - 2\,\text{Bi}^{*2} (3\,\text{Bi}^{*2} - 3\,\text{Bi}^* - 1)}{6 (r_f - \text{Bi}^*)^4} \right] \left(\frac{1}{\text{Bi}^*} - \frac{1}{r} \right)$$
$$- \frac{\text{Bi}^* r^2 - 3\,\text{Bi}^{*2} r + 6\,\text{Bi}^{*2} - 6\,\text{Bi}^* + 2}{6 r_f (r_f - \text{Bi}^*)^3}$$

where

$$\text{Bi}^* = \frac{\text{Bi}}{\text{Bi} - 1}$$

Discuss the behavior of the above solution as the freezing front approaches the center, that is, as $r_f \to 0$.

3.7 Utilize the expansions of Problem 3.6 in Eq. (3.81) and deduce the corresponding two-term solution for τ. Is this uniformly valid?

CHAPTER

FOUR

METHOD OF STRAINED COORDINATES

4.1 LIGHTHILL'S TECHNIQUE

When the systematic perturbation procedure leads to singular expansion, the result is of limited use unless it can be rendered uniformly valid. One of the main techniques for achieving uniform validity is the method of strained coordinates, which is the subject of this chapter.

The basic idea underlying the technique is to expand both the independent and dependent variables in terms of ϵ with coefficients expressed as functions of a new independent variable. The coefficients of the independent variable series are called the *straining functions.* The expansions are next substituted into the original equations to generate the usual sequence of perturbation equations. It is at this stage that the choice of the straining functions is made such that higher approximations are no more singular than the first. This principle is often referred to as Lighthill's rule, and its application is the crucial step of the whole analysis. If successful, the result is an implicit but uniformly valid solution. Because of its first appearance in Lighthill's paper (1949), the method is also called Lighthill's technique.

The spirit of Lighthill's technique is also reflected in earlier works of Lindstedt (1882) and Poincaré (1892) where, instead of a coordinate, a parameter is strained to achieve uniform validity. When a parameter is strained, the technique may be appropriately termed the method of strained parameters. Giving credit to the contributions of Poincaré, Lighthill, and later works of Kuo (1953, 1956), Tsien (1956) coined the name PLK method.

In the ensuing sections we illustrate the application of the technique to several problems.

4.1.1 A First-Order Differential Equation

Let us revert to the problem in Section 3.4 and concentrate on the series in Eq. (3.27) which exhibits nonuniformity in the neighborhood of $x = 0$. As noted there, the first term is singular at $x = 0$, and this singularity becomes worse in the subsequent terms. To remove the undesirable behavior, let us expand both y and x using a new independent variable s. Restricting ourselves to two-term expansions, we write

$$y = y_0(s) + \epsilon y_1(s) \tag{4.1}$$

$$x = s + \epsilon x_1(s) \tag{4.2}$$

where $x_1(s)$ is the straining function. Since we intend to strain x only slightly, we chose s itself as the leading term in Eq. (4.2). The choice of $x_1(s)$ is to be made such that Eqs. (4.1) and (4.2) give a uniformly valid solution.

Since

$$\frac{dy}{dx} = \frac{dy}{ds}\left(\frac{dx}{ds}\right)^{-1}$$

we have

$$\begin{aligned}\frac{dy}{dx} &= (y_0' + \epsilon y_1')(1 + \epsilon x_1')^{-1} \\ &= (y_0' + \epsilon y_1')(1 - \epsilon x_1' + \cdots) \\ &= y_0' + \epsilon(y_1' - y_0' x_1') + O(\epsilon^2)\end{aligned} \tag{4.3}$$

where primes denote differentiation with respect to s.

Substituting Eqs. (4.1)-(4.3) into Eq. (3.17) and equating coefficients of ϵ^0 and ϵ^1, we obtain

$$\epsilon^0: \quad s y_0' + y_0 = 0 \tag{4.4}$$

$$\epsilon^1: \quad s y_1' + y_1 = -(x_1 + y_0) y_0' + s y_0' x_1' \tag{4.5}$$

To derive the corresponding boundary conditions on y_0 and y_1, we proceed as follows. Let $\bar{s}$ be the value of s at $x = 1$; then from Eq. (4.2) it follows that

$$1 = \bar{s} + \epsilon x_1(\bar{s}) \tag{4.6}$$

Now assume a two-term expansion for $\bar{s}$ of the form

$$\bar{s} = 1 + \epsilon \bar{s}_1 \tag{4.7}$$

where $\bar{s}_1$ is to be determined. Substituting Eq. (4.7) into Eq. (4.6) gives

$$1 = 1 + \epsilon \bar{s}_1 + \epsilon x_1 (\bar{s} = 1 + \epsilon \bar{s}_1) \tag{4.8}$$

In Eq. (4.8) the parameter ϵ appears implicitly in the argument $\bar{s}(= 1 + \epsilon \bar{s}_1)$. To make Eq. (4.8) fully explicit in ϵ, we can expand the function x_1 about 1 to give

$$x_1 (\bar{s} = 1 + \epsilon \bar{s}_1) = x_1 (1) + \epsilon \bar{s}_1 x_1' (1) \tag{4.9}$$

Inserting Eq. (4.9) into Eq. (4.8) and retaining terms up to $O(\epsilon)$ gives

$$1 = 1 + \epsilon [\bar{s}_1 + x_1 (1)] \tag{4.10}$$

Thus it follows from Eq. (4.10) that

$$\bar{s}_1 = -x_1 (1) \tag{4.11}$$

and Eq. (4.7) now reads as

$$\bar{s} = 1 - \epsilon x_1 (1) \tag{4.12}$$

Using Eq. (4.12) in Eq. (4.1) and noting that $x = 1$ and $y = 1$, we get

$$1 = y_0 [\bar{s} = 1 - \epsilon x_1 (1)] + \epsilon y_1 [\bar{s} = 1 - \epsilon x_1 (1)] \tag{4.13}$$

in which ϵ again appears implicitly. Using Taylor expansions for y_0 and y_1 about 1, Eq. (4.13) becomes

$$1 = y_0 (1) + \epsilon [y_1 (1) - x_1 (1) y_0' (1)] + O(\epsilon^2) \tag{4.14}$$

Equating coefficients on both sides of Eq. (4.14), the boundary conditions are finally obtained as

$$y_0 (1) = 1 \qquad y_1 (1) = x_1 (1) y_0' (1) \tag{4.15}$$

We can now proceed with the solution of y_0 and y_1. The solution of Eq. (4.4) subject to Eq. (4.15) is

$$y_0 = \frac{1}{s} \tag{4.16}$$

Using Eq. (4.16) on the right-hand side of Eq. (4.5) gives

$$s y_1' + y_1 = \left(x_1 + \frac{1}{s} \right) \frac{1}{s^2} - \frac{1}{s} x_1' \tag{4.17}$$

At this stage we tackle the problem of choosing the straining function x_1. The choice is dictated by the principle that the solution for y_1 should be no more singular than the solution for y_0. We discuss two alternative approaches.

Approach 1. One way to meet Lighthill's principle is to solve for y_1 making sure that it is no more singular than y_0. To facilitate integration, Eq. (4.17) is rewritten as

$$(sy_1)' = -\left(\frac{x_1}{s} + \frac{1}{2s^2}\right)' \tag{4.18}$$

which readily integrates to give

$$y_1 = \frac{C}{s} - \frac{1}{s^2}\left(x_1 + \frac{1}{2s}\right) \tag{4.19}$$

where C is a constant of integration. To ensure that y_1 does not contain powers of s higher than unity in the denominator, we must put

$$\frac{x_1}{s} + \frac{1}{2s^2} = \text{constant or regular function of } s \tag{4.20}$$

The simplest choice is to use zero on the right-hand side of Eq. (4.20). This gives

$$x_1 = -\frac{1}{2s} \tag{4.21}$$

Introducing Eq. (4.21) into Eq. (4.19) and applying the boundary condition in Eq. (4.15), which gives $y_1(1) = \frac{1}{2}$, the solution for y_1 becomes

$$y_1 = \frac{1}{2s} \tag{4.22}$$

Thus the final result is

$$y = \frac{1}{s} + \epsilon \frac{1}{2s} \tag{4.23}$$

$$x = s - \epsilon \frac{1}{2s} \tag{4.24}$$

which constitute a uniformly valid solution. However, the presence of the variable s makes the solution implicit. This implicitness is an essential feature of the technique, but in this particular case, an explicit solution can be derived by eliminating s. From Eq. (4.23) $s = (2 + \epsilon)/2y$, which can be used in Eq. (4.24) to obtain a quadratic in y

$$\frac{2\epsilon}{2+\epsilon} y^2 + 2xy - (2 + \epsilon) = 0 \tag{4.25}$$

whose solution is

$$y = \frac{2+\epsilon}{2\epsilon}\left(-x + \sqrt{x^2 + 2\epsilon}\right) \tag{4.26}$$

As $x \to 0, y \to (2+\epsilon)/\sqrt{2\epsilon}$ and hence the solution is uniformly valid.

We could also have dropped the second term in y but retained two terms in x in which case the explicit solution would be

$$y = -\frac{x}{\epsilon} + \sqrt{\left(\frac{x}{\epsilon}\right)^2 + \frac{2}{\epsilon}} \tag{4.27}$$

As $x \to 0, y \to \sqrt{2/\epsilon}$ and the solution is again uniformly valid. Although both Eqs. (4.26) and (4.27) are uniformly valid, they are not equally accurate. For example, at $\epsilon = 0.1$, the values of $y(0)$ from Eqs. (4.26) and (4.27) are 4.696 and 4.472, respectively, whereas the exact value is 4.583. Figure 4.1 compares Eqs. (4.26) and (4.27) with the exact solution in Eq. (3.29) for $\epsilon = 0.1$.

Approach 2. An alternative way to choose x_1 is to simply examine the equation for y_1 (not solve it). To ensure that y_1 be no more singular than y_0, we can set the right-hand side of Eq. (4.18) equal to zero giving a differential equation for x_1 as

$$\left(\frac{x_1}{s} + \frac{1}{2s^2}\right)' = 0 \tag{4.28}$$

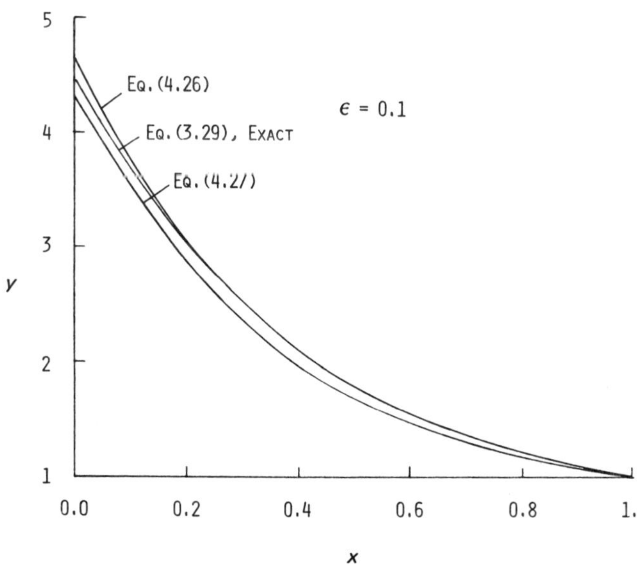

Figure 4.1 Strained coordinate and exact solutions of Eqs. (3.17) and (3.18).

Obviously, to solve Eq. (4.28), a boundary condition is needed. Since the perturbation expansion in Section 3.4 is satisfactory in the neighborhood of $x = 1$ (see Fig. 3.1), we stipulate that the straining should gradually diminish in this region and ultimately vanish at $x = 1$. Thus we impose the condition

$$x_1(1) = 0 \tag{4.29}$$

so that according to Eq. (4.2), $x = 1$ gives $s = 1$ and the straining vanishes.

The solution of Eq. (4.28) subject to Eq. (4.29) is

$$x_1 = \frac{s^2 - 1}{2s} \tag{4.30}$$

Thus an alternative solution is

$$y = \frac{1}{s} \tag{4.31}$$

$$x = s + \epsilon \frac{s^2 - 1}{2s} \tag{4.32}$$

which again is implicit. Since Eq. (4.32) is a quadratic in s, we can derive an explicit solution. When this is done, one recovers, rather fortuitously, the exact solution which is Eq. (3.29). Now that Eqs. (4.31) and (4.32) represent the exact solution, all higher-order terms must vanish. For verification let us consider the solution for y_1. In view of Eq. (4.28), Eq. (4.18) reduces to

$$(sy_1)' = 0 \tag{4.33}$$

and the boundary condition in Eq. (4.15) becomes

$$y_1(1) = 0 \tag{4.34}$$

From Eqs. (4.33) and (4.34) it is obvious that y_1 is identically zero.

Comparing the two alternative solutions, it is interesting to note that in Eqs. (4.23) and (4.24) the straining does not vanish at $x = 1$. Being exact, the solution in Eqs. (4.31) and (4.32) is obviously superior to the solution in Eqs. (4.23) and (4.24).

4.1.2 Melting of a Finite Slab

In Section 3.6 it was noted that the solution for v_2 exhibited erratic behavior due to the presence of terms $\tau e^{-3\tau}$ and $\tau^2 e^{-3\tau}$. To eliminate the influence of these undesirable terms, we use the method of strained coordinates. Introducing a new independent variable s, we expand τ as

$$\tau = s + \epsilon\tau_1(s) + \epsilon^2\tau_2(s) \tag{4.35}$$

where $\tau_1(s)$ and $\tau_2(s)$ are the straining functions. Now

$$\frac{du}{d\tau}=\frac{du}{ds}\left(\frac{d\tau}{ds}\right)^{-1}=u_0'+\epsilon(u_1'-\tau_1'u_0')+\epsilon^2[(\tau_1'^2-\tau_2')u_0'-\tau_1'u_1'+u_2']$$
$$+O(\epsilon^3) \tag{4.36}$$

$$\frac{dv}{d\tau}=\frac{dv}{ds}\left(\frac{d\tau}{ds}\right)^{-1}=v_0'+\epsilon(v_1'-\tau_1'v_0')+\epsilon^2[(\tau_1'^2-\tau_2')v_0'-\tau_1'v_1'+v_2']$$
$$+O(\epsilon^3) \tag{4.37}$$

$$\frac{dw}{d\tau}=\frac{dw}{ds}\left(\frac{d\tau}{ds}\right)^{-1}=w_0'+\epsilon(w_1'-\tau_1'w_0')+\epsilon^2[(\tau_1'^2-\tau_2')w_0'-\tau_1'w_1'+w_2']$$
$$+O(\epsilon^3) \tag{4.38}$$

where primes denote differentiation with respect to s.

Since the perturbation expansion of Section 3.6 is accurate in the neighborhood of $\tau = 0$, we make the straining vanish at $\tau = 0$, i.e.,

$$\tau_1(0)=\tau_2(0)=0 \tag{4.39}$$

which gives the initial condition in terms of s as

$$s=0 \quad u=0 \quad v=-\tfrac{2}{3}\epsilon \quad w=0 \tag{4.40}$$

Substituting Eqs. (3.51)-(3.53) and Eqs. (4.36)-(4.40) into Eqs. (3.47)-(3.50) and equating coefficients of ϵ^0, ϵ^1, and ϵ^2, we obtain

$$\epsilon^0:\quad u_0'+\mu v_0'+\mu w_0'=0 \tag{4.41}$$

$$(1-u_0)^2v_0'=-3v_0 \tag{4.42}$$

$$\tfrac{1}{3}u_0^2w_0'=-v_0w_0 \tag{4.43}$$

$$s=0 \quad u=v=w=0 \tag{4.44}$$

$$\epsilon^1:\quad u_1'-\tau_1'u_0'+\mu(v_1'-\tau_1'v_0')+\mu(w_1'-\tau_1'w_0')=2\mu \tag{4.45}$$

$$(v_1'-\tau_1'v_0')(1-u_0)^2-2u_1(1-u_0)v_0'=-3v_1 \tag{4.46}$$

$$\tfrac{1}{3}(w_1'-\tau_1'w_0')u_0^2+\tfrac{2}{3}u_0u_1w_0'=u_0^2-(v_0w_1+w_0v_1) \tag{4.47}$$

$$s=0 \quad u_1=w_1=0 \quad v_1=-\tfrac{2}{3} \tag{4.48}$$

$$\epsilon^2:\quad (\tau_1'^2-\tau_2')u_0'-\tau_1'u_1'+u_2'+\mu[(\tau_1'^2-\tau_2')v_0'-\tau_1'v_1'+v_2']$$
$$+\mu[(\tau_1'^2-\tau_2')w_0'-\tau_1'w_1'+w_2']=0 \tag{4.49}$$

$$[(\tau_1'^2-\tau_2')v_0'-\tau_1'v_1'+v_2'](1-u_0)^2-2(v_1'-\tau_1'v_0')u_1(1-u_0)$$
$$+(2u_0u_2+u_1^2-2u_2)v_0'=-3v_2 \tag{4.50}$$

$$\tfrac{1}{3}\{[(\tau_1'^2 - \tau_2')w_0' - \tau_1' w_1' + w_2']u_0^2 + 2(w_1' - \tau_1' w_0')u_0 u_1 + (2u_0 u_2 + u_1^2)w_0'\} = 2u_0 u_1 - (v_0 w_2 + v_1 w_1 + w_0 v_2) \tag{4.51}$$

$$s = 0 \qquad u_2 = v_2 = w_2 = 0 \tag{4.52}$$

The solutions of Eqs. (4.41)-(4.48) are readily seen to be the same as those derived in Section 3.6, that is,

$$u_0 = v_0 = w_0 = 0 \tag{4.53}$$

$$w_1 = 0 \tag{4.54}$$

$$v_1 = -\tfrac{2}{3}\mu e^{-3s} \tag{4.55}$$

$$u_1 = \tfrac{2}{3}\mu(e^{-3s} + 3s + 1) \tag{4.56}$$

The second-order equations [Eqs. (4.49)-(4.51)] now reduce to

$$w_2 = 0 \tag{4.57}$$

$$v_2' + 3v_2 = 2e^{-3s}[\tau_1' + 4\mu(s - \tfrac{1}{3}) + \tfrac{4}{3}\mu e^{-3s}] \tag{4.58}$$

$$u_2' + \mu v_2' + \mu w_2' = 2\mu\tau_1' \tag{4.59}$$

Examining Eq. (4.58) it is evident that any terms containing e^{-3s} will give rise to undesirable terms. To avoid such terms, we set the coefficient of e^{-3s} to zero giving

$$\tau_1' + 4\mu(s - \tfrac{1}{3}) = 0 \tag{4.60}$$

The solution of Eq. (4.60) subject to the initial condition given in Eq. (4.39) is

$$\tau_1 = -\tfrac{2}{3}\mu s(3s - 2) \tag{4.61}$$

The solutions for v_2 and u_2 now follow as

$$v_2 = \tfrac{8}{9}\mu(e^{-3s} - e^{-6s}) \tag{4.62}$$

$$u_2 = -\tfrac{4}{9}\mu^2(9s^2 - 6s + 2e^{-3s} - 2e^{-6s}) \tag{4.63}$$

Thus, a uniformly valid solution is

$$u = \tfrac{2}{3}\epsilon\mu(e^{-3s} + 3s - 1) - \tfrac{4}{9}\epsilon^2\mu^2(9s^2 - 6s + 2e^{-3s} - 2e^{-6s}) + O(\epsilon^3) \tag{4.64}$$

$$v = -\tfrac{2}{3}\epsilon e^{-3s} + \tfrac{8}{9}\epsilon^2\mu(e^{-3s} - e^{-6s}) + O(\epsilon^3) \tag{4.65}$$

$$w = 0 \tag{4.66}$$

$$\tau = s - \tfrac{2}{3}\epsilon\mu s(3s - 2) + O(\epsilon^2) \tag{4.67}$$

4.1.3 Inward Spherical Solidification

Unlike the previous two examples, the governing differential equation [Eq. (3.79)] for inward spherical solidification is a partial differential equation involving two independent variables r and r_f. From the analysis given in Section 3.7 we recall that the solutions for u_1 and u_2 are singular as $r_f \to 0$. Similarly, the second-order term in the series [Eq. (3.99)] for τ becomes singular as $r_f \to 0$.

To obtain a uniformly valid solution of Eqs. (3.79) and (3.80), we follow Pedroso and Domoto (1973c), introduce two new variables ϕ and ψ, and expand u, r, and r_f as

$$u = u_0(\phi, \psi) + \epsilon u_1(\phi, \psi) + \epsilon^2 u_2(\phi, \psi) \tag{4.68}$$

$$r = \phi + \epsilon\sigma_1(\phi, \psi) + \epsilon^2\sigma_2(\phi, \psi) \tag{4.69}$$

$$r_f = \psi + \epsilon\sigma_1(\psi, \psi) + \epsilon^2\sigma_2(\psi, \psi) \tag{4.70}$$

where the straining functions $\sigma_1(\phi, \psi)$ and $\sigma_2(\phi, \psi)$ will be chosen to ensure uniform validity of the solution. Note that as $r \to r_f$, $\phi \to \psi$. Since the regular perturbation solution is satisfactory near the container boundary, we choose to make the straining vanish at the boundary so that

$$\lim_{\phi \to 1} \sigma_i(\phi, \psi) = 0 \qquad (r \to 1, \phi \to 1) \tag{4.71}$$

$$\lim_{\psi \to 1} \sigma_i(\psi, \psi) = 0 \qquad (r_f \to 1, \psi \to 1) \tag{4.72}$$

Also note that r is a function of both ϕ and ψ but r_f is a function of ψ alone.

To change the variables from (r, r_f) to (ϕ, ψ) in Eqs. (3.79) and (3.80) we proceed as follows. For any function $f(r, r_f)$ where $r = f_1(\phi, \psi)$ and $r_f = f_2(\psi)$, we have

$$\frac{\partial f}{\partial \phi} = \frac{\partial f}{\partial r}\frac{\partial r}{\partial \phi} + \frac{\partial f}{\partial r_f}\frac{\partial r_f}{\partial \phi} \tag{4.73}$$

$$\frac{\partial f}{\partial \psi} = \frac{\partial f}{\partial r}\frac{\partial r}{\partial \psi} + \frac{\partial f}{\partial r_f}\frac{\partial r_f}{\partial \psi} \tag{4.74}$$

Since $r_f = f_2(\psi)$, $\partial r_f/\partial\phi = 0$; hence

$$\frac{\partial f}{\partial \phi} = \frac{\partial f}{\partial r}\frac{\partial r}{\partial \phi} \tag{4.75}$$

From Eqs. (4.75) and (4.74) we have

$$\frac{\partial f}{\partial r} = \frac{\partial f}{\partial \phi}\left(\frac{\partial r}{\partial \phi}\right)^{-1} \tag{4.76}$$

$$\frac{\partial f}{\partial r_f} = \left(\frac{\partial f}{\partial \psi} - \frac{\partial f}{\partial r}\frac{\partial r}{\partial \psi}\right)\left(\frac{\partial r_f}{\partial \psi}\right)^{-1}$$

or

$$\frac{\partial f}{\partial r_f} = \left[\frac{\partial f}{\partial \psi} - \frac{\partial f}{\partial \phi}\left(\frac{\partial r}{\partial \phi}\right)^{-1}\frac{\partial r}{\partial \psi}\right]\left(\frac{\partial r_f}{\partial \psi}\right)^{-1} \tag{4.77}$$

Since we need the second derivative with respect to r, we differentiate Eq. (4.76) to get

$$\frac{\partial^2 f}{\partial r^2} = \left[\frac{\partial^2 f}{\partial \phi^2} - \frac{\partial f}{\partial \phi}\frac{\partial^2 r}{\partial \phi^2}\left(\frac{\partial r}{\partial \phi}\right)^{-1}\right]\left(\frac{\partial r}{\partial \phi}\right)^{-2} \tag{4.78}$$

For Eq. (3.79) the quantities required are $\partial u/\partial r$, $\partial u/\partial r_f$, and $\partial(ur)/\partial r^2$. Based on Eqs. (4.76)-(4.78), we can write

$$\frac{\partial u}{\partial r} = \left(\frac{\partial u}{\partial \phi}\right)\left(\frac{\partial r}{\partial \phi}\right)^{-1} \tag{4.79}$$

$$\frac{\partial u}{\partial r_f} = \left[\frac{\partial u}{\partial \psi} - \frac{\partial u}{\partial \phi}\left(\frac{\partial r}{\partial \phi}\right)^{-1}\frac{\partial r}{\partial \psi}\right]\left(\frac{\partial r_f}{\partial \psi}\right)^{-1} \tag{4.80}$$

$$\frac{\partial^2 (ru)}{\partial r^2} = \left[\frac{\partial^2 (ru)}{\partial \phi^2} - \frac{\partial (ru)}{\partial \phi}\frac{\partial^2 r}{\partial \phi^2}\left(\frac{\partial r}{\partial \phi}\right)^{-1}\right]\left(\frac{\partial r}{\partial \phi}\right)^{-2} \tag{4.81}$$

Using Eqs. (4.68)-(4.70), we derive the following derivatives

$$\frac{\partial u}{\partial \phi} = u_{0\phi} + \epsilon u_{1\phi} + \epsilon^2 u_{2\phi} \tag{4.82}$$

$$\frac{\partial u}{\partial \psi} = u_{0\psi} + \epsilon u_{1\psi} + \epsilon^2 u_{2\psi} \tag{4.83}$$

$$\frac{\partial r}{\partial \phi} = 1 + \epsilon \sigma_{1\phi} + \epsilon^2 \sigma_{2\phi} \tag{4.84}$$

$$\frac{\partial^2 r}{\partial \phi^2} = \epsilon \sigma_{1\phi\phi} + \epsilon^2 \sigma_{2\phi\phi} \tag{4.85}$$

$$\frac{\partial r}{\partial \psi} = \epsilon \sigma_{1\psi} + \epsilon^2 \sigma_{2\psi} \tag{4.86}$$

$$\frac{\partial r_f}{\partial \psi} = 1 + \epsilon \hat{\sigma}_{1\psi} + \epsilon^2 \hat{\sigma}_{2\psi} \tag{4.87}$$

where the subscripts ϕ and ψ are used from now on to indicate partial derivatives, and

$$\begin{aligned} \sigma_1 &= \sigma_1(\phi, \psi) \\ \sigma_2 &= \sigma_2(\phi, \psi) \\ \hat{\sigma}_1 &= \sigma_1(\psi, \psi) \\ \hat{\sigma}_2 &= \sigma_2(\psi, \psi) \end{aligned} \tag{4.88}$$

Additionally, we need the derivatives $\partial(ru)/\partial\phi$ and $\partial^2(ru)/\partial\phi^2$. From Eqs. (4.68) and (4.69), we have

$$ru = (\phi + \epsilon\sigma_1 + \epsilon^2\sigma_2)(u_0 + \epsilon u_1 + \epsilon^2 u_2)$$

or

$$ru = \phi u_0 + \epsilon(u_0\sigma_1 + \phi u_1) + \epsilon^2(u_0\sigma_2 + u_1\sigma_1 + \phi u_2)$$

Thus

$$\frac{\partial(ru)}{\partial\phi} = (\phi u_0)_\phi + \epsilon(u_0\sigma_1 + \phi u_1)_\phi + \epsilon^2(u_0\sigma_2 + u_1\sigma_1 + \phi u_2)_\phi \qquad (4.89)$$

and

$$\frac{\partial^2(ru)}{\partial\phi^2} = (\phi u_0)_{\phi\phi} + \epsilon(u_0\sigma_1 + \phi u_1)_{\phi\phi} + \epsilon^2(u_0\sigma_2 + u_1\sigma_1 + \phi u_2)_{\phi\phi} \qquad (4.90)$$

Returning to Eq. (4.79), we have

$$\frac{\partial u}{\partial r} = (u_{0\phi} + \epsilon u_{1\phi} + \epsilon^2 u_{2\phi})(1 + \epsilon\sigma_{1\phi} + \epsilon^2\sigma_{2\phi})^{-1}$$

or

$$\frac{\partial u}{\partial r} = u_{0\phi} + \epsilon(u_{1\phi} - u_{0\phi}\sigma_{1\phi}) \qquad (4.91)$$

where we have retained only terms up to $O(\epsilon)$. Since ϵ appears on the right-hand side of Eq. (3.79), we need to retain only two terms in Eq. (4.91) to obtain the solution up to $O(\epsilon^2)$.

We now work out the expansion for $\partial u/\partial r_f$. Using Eq. (4.80), and retaining terms only up to $O(\epsilon)$, we get

$$\frac{\partial u}{\partial r_f} = [(u_{0\psi} + \epsilon u_{1\psi}) - (u_{0\phi} + \epsilon u_{1\phi})(1 \quad \epsilon\sigma_{1\phi})(\epsilon\sigma_{1\psi})](1 - \epsilon\hat{\sigma}_{1\psi})$$

or

$$\frac{\partial u}{\partial r_f} = u_{0\psi} + \epsilon(u_{1\psi} - u_{0\phi}\sigma_{1\psi} - u_{0\psi}\hat{\sigma}_{1\psi}) \qquad (4.92)$$

Finally, coming to Eq. (4.81), we have

$$\begin{aligned}\frac{\partial^2(ru)}{\partial r^2} = {} & \{(\phi u_0)_{\phi\phi} + \epsilon(u_0\sigma_1 + \phi u_1)_{\phi\phi} + \epsilon^2(u_0\sigma_2 + u_1\sigma_1 + \phi u_2)_{\phi\phi} \\ & - [(\phi u_0)_\phi + \epsilon(u_0\sigma_1 + \phi u_1)_\phi + \epsilon^2(u_0\sigma_2 + u_1\sigma_1 + \phi u_2)_\phi] \\ & \times (\epsilon\sigma_{1\phi\phi} + \epsilon^2\sigma_{2\phi\phi})[1 - \epsilon\sigma_{1\phi} + \epsilon^2(\sigma_{1\phi}^2 - \sigma_{2\phi})]\} \\ & \times [1 - \epsilon 2\sigma_{1\phi} + \epsilon^2(3\sigma_{1\phi}^2 - 2\sigma_{2\phi})]\end{aligned}$$

or

$$\frac{\partial^2 (ru)}{\partial r^2} = (\phi u_0)_{\phi\phi} + \epsilon[(\phi u_1 + u_0\sigma_1)_{\phi\phi} - \sigma_{1\phi\phi}(\phi u_0)_\phi - 2\sigma_{1\phi}(\phi u_0)_{\phi\phi}]$$
$$+ \epsilon^2 [(\phi u_2 + u_1\sigma_1 + u_0\sigma_2)_{\phi\phi} - \sigma_{1\phi\phi}(\phi u_1 + u_0\sigma_1)_\phi$$
$$- 2\sigma_{1\phi}(\phi u_1 + u_0\sigma_1)_{\phi\phi} - (2\sigma_{2\phi} - 3\sigma_{1\phi}^2)(\phi u_0)_{\phi\phi}$$
$$- (\sigma_{2\phi\phi} - 3\sigma_{1\phi}\sigma_{1\phi\phi})(\phi u_0)_\phi] \tag{4.93}$$

The right-hand side of Eq. (3.79), when multiplied by r, can now be written as

$$\epsilon r \frac{\partial u}{\partial r}\bigg|_{r=r_f} \frac{\partial u}{\partial r_f} = \epsilon(\phi + \epsilon\sigma_1)[u_{0\phi}|_{\phi=\psi} + \epsilon(u_{1\phi} - u_{0\phi}\sigma_{1\phi})|_{\phi=\psi}]$$
$$\times [u_{0\psi} + \epsilon(u_{1\psi} - u_{0\phi}\sigma_{1\psi} - u_{0\psi}\hat{\sigma}_{1\psi})]$$
$$= \epsilon\{\phi u_{0\phi}|_{\phi=\psi} u_{0\psi} + \epsilon[\phi u_{0\phi}|_{\phi=\psi}(u_{1\psi} - u_{0\phi}\sigma_{1\psi} - u_{0\psi}\hat{\sigma}_{1\psi})$$
$$+ u_{0\psi}(\sigma_1 u_{0\phi}|_{\phi=\psi} + (u_{1\phi} - u_{0\phi}\sigma_{1\phi})|_{\phi=\psi}\,\phi)]\} \tag{4.94}$$

We now use Eqs. (4.93) and (4.94) to write the sequence of perturbation equations as follows

$$\epsilon^0: \quad (\phi u_0)_{\phi\phi} = 0 \tag{4.95}$$

$$u_0(\psi, \phi = 1) = 0 \qquad u_0(\psi, \phi = \psi) = 1 \tag{4.96}$$

$$\epsilon^1: \quad (\phi u_1 + u_0\sigma_1)_{\phi\phi} - \sigma_{1\phi\phi}(\phi u_0)_\phi = \phi u_{0\phi}|_{\phi=\psi} u_{0\psi} \tag{4.97}$$

$$u_1(\psi, \phi = 1) = 0 \qquad u_1(\psi, \phi = \psi) = 0 \tag{4.98}$$

$$\epsilon^2: \quad (\phi u_2 + u_1\sigma_1 + u_0\sigma_2)_{\phi\phi} - \sigma_{1\phi\phi}(\phi u_1 + u_0\sigma_1)_\phi - 2\sigma_{1\phi}(\phi u_1 + u_0\sigma_1)_{\phi\phi}$$
$$- (\sigma_{2\phi\phi} - 3\sigma_{1\phi}\sigma_{1\phi\phi})(\phi u_0)_\phi = \phi u_{0\phi}|_{\phi=\psi}(u_{1\psi} - u_{0\phi}\sigma_{1\psi}$$
$$- u_{0\psi}\hat{\sigma}_{1\psi}) + u_{0\psi}[\sigma_1 u_{0\phi}|_{\phi=\psi} + \phi(u_{1\phi} - u_{0\phi}\sigma_{1\rho})_{\phi=\psi}] \tag{4.99}$$

$$u_2(\psi, \phi = 1) = 0 \qquad u_2(\psi, \phi = \psi) = 0 \tag{4.100}$$

Zero-order solution. Since Eqs. (4.95) and (4.96) have the same form as Eqs. (3.83) and (3.84), the solution is

$$u_0 = \frac{1 - 1/\phi}{1 - 1/\psi} \tag{4.101}$$

First-order solution. From Eq. (4.101) we have

$$u_{0\phi} = \frac{\psi}{\phi^2(\psi - 1)} \qquad u_{0\phi}|_{\phi=\psi} = -\frac{1}{\psi(1-\psi)}$$
$$\phi u_0 = \frac{\psi(\phi - 1)}{\psi - 1} \qquad (\phi u_0)_\phi = -\frac{\psi}{1 - \psi} \qquad u_{0\psi} = \frac{1 - \phi}{\phi(1 - \psi)^2} \tag{4.102}$$

Using Eqs. (4.101) and (4.102), Eq. (4.97) can be simplified to

$$\left(\phi u_1 + \frac{\psi}{1-\psi}\frac{\sigma_1}{\phi}\right)_{\phi\phi} = -\frac{1-\phi}{\psi(1-\psi)^3} \tag{4.103}$$

Integrating Eq. (4.103), we have

$$\left(\phi u_1 + \frac{\psi}{1-\psi}\frac{\sigma_1}{\phi}\right)_{\phi} = -\frac{\phi(2-\phi)}{2\psi(1-\psi)^3} + f_1(\psi)$$

Let u_1 be *identically zero*. Then, imposing the condition $\lim_{\phi\to 1}(\sigma_1/\phi)_\phi = 0$ gives

$$f_1(\psi) = \frac{1}{2\psi(1-\psi)^3}$$

Integrating once again, we have

$$\frac{\psi}{1-\psi}\frac{\sigma_1}{\phi} = -\frac{\phi^2(3-\phi)}{6\psi(1-\psi)^3} + \frac{\phi}{2\psi(1-\psi)^3} + f_2(\psi)$$

Using the condition in Eq. (4.71) gives

$$f_2(\psi) = -\frac{1}{6\psi(1-\psi)^3}$$

Hence the solution for σ_1 becomes

$$\sigma_1 = -\frac{\phi(1-\phi)^3}{6\psi^2(1-\psi)^2} \tag{4.104}$$

Second-order solution. By making u_2 *identically zero*, Eq. (4.99) reduces to

$$\begin{aligned}(u_0\sigma_2)_{\phi\phi} &- \sigma_{1\phi\phi}(u_0\sigma_1)_\phi - 2\sigma_{1\phi}(u_0\sigma_1)_{\phi\phi} - (\sigma_{2\phi\phi} - 3\sigma_{1\phi}\sigma_{1\phi\phi})(\phi u_0)_\phi \\ &= -\phi u_{0\phi}|_{\phi=\psi}(u_{0\phi}\sigma_{1\psi} + u_{0\psi}\hat{\sigma}_{1\psi}) + u_{0\psi}[\sigma_1 u_{0\phi}|_{\phi=\psi} \\ &\quad - \phi u_{0\phi}|_{\phi=\psi}\sigma_{1\phi}|_{\phi=\psi}]\end{aligned} \tag{4.105}$$

We now evaluate the various quantities appearing in Eq. (4.105). From Eq. (4.104), we have

$$\sigma_{1\phi} = \frac{(1-\phi)^2(4\phi-1)}{6\psi^2(1-\psi)^2} \qquad \sigma_{1\phi\phi} = \frac{(1-\phi)(1-2\phi)}{\psi^2(1-\psi)^2}$$

From Eqs. (4.101) and (4.104) it follows that

$$u_0\sigma_1 = -\frac{(1-\phi)^4}{6\psi(1-\psi)^3}$$

and

$$(u_0\sigma_1)_\phi = \frac{2(1-\phi)^3}{3\psi(1-\psi)^3} \qquad (u_0\sigma_1)_{\phi\phi} = -\frac{2(1-\phi)^2}{\psi(1-\psi)^3}$$

Also, from Eq. (4.104) we obtain

$$\hat{\sigma}_1 = \sigma_1(\psi, \psi) = -\frac{1-\psi}{6\psi}$$

$$\sigma_{1\psi} = \frac{\phi(1-\phi)^3(1-2\psi)}{3\psi^3(1-\psi)^3}$$

and

$$\hat{\sigma}_{1\psi} = \frac{1}{6\psi^2}$$

Using the foregoing information we can evaluate the various terms appearing in Eq. (4.105) as follows

$$\phi_{1\phi\phi}(u_0\sigma_1)_\phi = \frac{2(1-2\phi)(1-\phi)^4}{3\psi^3(1-\psi)^5}$$

$$2\sigma_{1\phi}(u_0\sigma_1)_{\phi\phi} = -\frac{2(4\phi-1)(1-\phi)^4}{3\psi^3(1-\psi)^5}$$

$$3\sigma_{1\phi}\sigma_{1\phi\phi}(\phi u_0)_\phi = \frac{(8\phi^2-6\phi+1)(1-\phi)^3}{2\psi^3(1-\psi)^5}$$

$$\phi u_{0\phi}|_{\phi=\psi}(u_{0\phi}\sigma_{1\psi} + u_{0\psi}\hat{\sigma}_{1\psi}) = \frac{(2-4\psi)(1-\phi)^3}{6\psi^3(1-\psi)^5} - \frac{1-\phi}{6\psi^3(1-\psi)^3}$$

$$u_{0\psi}(\sigma_1 u_{0\phi}|_{\phi=\psi} - \phi u_{0\phi}|_{\phi=\psi}\sigma_{1\phi}|_{\phi=\psi}) = \frac{(1-\phi)^4}{6\psi^3(1-\psi)^5} + \frac{(1-\phi)(4\psi-1)}{6\psi^3(1-\psi)^3}$$

Making use of the foregoing information and simplifying, Eq. (4.105) now appears as

$$\frac{\psi}{1-\psi}\left(\frac{\sigma_2}{\phi}\right)_{\phi\phi} = \frac{(4+4\psi-15\phi)(1-\phi)^3}{6\psi^3(1-\psi)^5} + \frac{2(1-\phi)}{3\psi^2(1-\psi)^3} \tag{4.106}$$

Integrating Eq. (4.106), we get

$$\frac{\psi}{1-\psi}\left(\frac{\sigma^2}{\phi}\right)_\phi = \frac{(1-\phi)^4(12\phi-4\psi-1)}{24\psi^3(1-\psi)^5} + \frac{\phi(2-\phi)}{3\psi^2(1-\psi)^3} + f_3(\psi)$$

Imposing the condition $\lim_{\phi\to 1}(\sigma_2/\phi)_\phi = 0$ gives

$$f_3(\psi) = -\frac{1}{3\psi^2(1-\psi)^3}$$

Integrating once again, we get

$$\frac{\psi}{1-\psi}\frac{\sigma_2}{\phi} = -\frac{(1-\phi)^5(1-4\psi+10\phi)}{120\psi^3(1-\psi)^5} + \frac{\phi^2(3-\phi)}{9\psi^2(1-\psi)^3} - \frac{\phi}{3\psi^2(1-\psi)^3} + f_4(\psi)$$

Using the condition in Eq. (4.71) gives

$$f_4(\psi) = \frac{1}{9\psi^2(1-\psi)^3}$$

Thus the solution for σ_2 becomes

$$\sigma_2 = \frac{\phi(1-\phi)^3}{\psi^3(1-\psi)^2}\left[\frac{1}{9} - \frac{(1-\phi)^2(1-4\psi+10\phi)}{120\psi(1-\psi)^2}\right] \tag{4.107}$$

From Eq. (4.107) it follows immediately that

$$\hat{\sigma}_2 = \sigma_2(\psi, \psi) = \frac{(22\psi-3)(1-\psi)}{360\psi^3}$$

Also, for subsequent use, we deduce

$$\sigma_{2\phi}|_{\phi=\psi} = \frac{-22\psi^2+10\psi-3}{360\psi^4} \qquad \hat{\sigma}_{2\psi} = \frac{22\psi^2-50\psi+9}{360\psi^4}$$

Having determined the straining functions, we can now write the final solution from Eqs. (4.68)-(4.70) as

$$u = \frac{1-1/\phi}{1-1/\psi} + O(\epsilon^3) \tag{4.108}$$

$$r = \phi - \epsilon\frac{\phi(1-\phi)^3}{6\psi^2(1-\psi)^2} + \epsilon^2\frac{\phi(1-\phi)^3}{\psi^3(1-\psi)^2}\left[\frac{1}{9} - \frac{(1-\phi)^2(1+4\psi+10\phi)}{120\psi(1-\psi)^2}\right] \tag{4.109}$$

$$r_f = \psi - \epsilon\frac{1-\psi}{6\psi} + \epsilon^2\frac{(22\psi-3)(1-\psi)}{360\psi^3} \tag{4.110}$$

Freezing time. To obtain the freezing time τ as a function of ψ, we consider Eq. (3.81). Since

$$\frac{dr_f}{d\tau} = \frac{dr_f}{d\psi}\left(\frac{d\tau}{d\psi}\right)^{-1}$$

we have, using Eq. (3.81),

$$\frac{d\tau}{d\psi} = \left(\left.\frac{\partial u}{\partial r}\right|_{r=r_f}\right)^{-1}\frac{dr_f}{d\psi} \tag{4.111}$$

To obtain the expansion for $(\partial u/\partial r)|_{r=r_f}$ up to $O(\epsilon^2)$, we return to Eq. (4.79). Noting that u_1 and u_2 are identically zero, we have

$$\begin{aligned}\frac{\partial u}{\partial r} &= u_{0\phi}(1+\epsilon\sigma_{1\phi}+\epsilon^2\sigma_{2\phi})^{-1}\\ &= u_{0\phi} - \epsilon\sigma_{1\phi}u_{0\phi} - \epsilon^2[(\sigma_{2\phi}-\sigma_{1\phi}^2)u_{0\phi}] + O(\epsilon^3)\end{aligned} \tag{4.112}$$

Hence

$$\left.\frac{\partial u}{\partial r}\right|_{r=r_f} = u_{0\phi}|_{\phi=\psi} - \epsilon(\sigma_{1\phi}u_{0\phi})_{\phi=\psi} - \epsilon^2\,[(\sigma_{2\phi} - \sigma_{1\phi}^2)u_{0\phi}]_{\phi=\psi} + O(\epsilon^3) \tag{4.113}$$

Substituting Eq. (4.113) into Eq. (4.111) we have

$$\frac{d\tau}{d\psi} = \{u_{0\phi}|_{\phi=\psi} - \epsilon(\sigma_{1\phi}u_{0\phi})_{\phi=\psi} - \epsilon^2\,[(\sigma_{2\phi} - \sigma_{1\phi}^2)u_{0\phi}]_{\phi=\psi}\}^{-1}$$
$$\times (1 + \epsilon\hat{\sigma}_{1\psi} + \epsilon^2\hat{\sigma}_{2\psi})$$

which can be expanded and simplified to give

$$\frac{d\tau}{d\psi} = (u_{0\phi}|_{\phi=\psi})^{-1}\,[1 + \epsilon(\sigma_{1\phi}|_{\phi=\psi} + \hat{\sigma}_{1\psi})$$
$$+ \epsilon^2(\hat{\sigma}_{2\psi} + \sigma_{1\phi}|_{\phi=\psi}\hat{\sigma}_{1\psi} + \sigma_{2\phi}|_{\phi=\psi})] \tag{4.114}$$

Utilizing the appropriate expressions for quantities appearing in Eq. (4.114), it is found that

$$\frac{d\tau}{d\psi} = -\psi(1-\psi) - \epsilon\frac{2}{3}(1-\psi) + \epsilon^2\frac{1}{90}\left(\frac{1}{\psi^3} - \frac{1}{\psi^2}\right) \tag{4.115}$$

Integrating Eq. (4.115) and imposing the condition $\psi = 1$, $\tau = 0$, the final result is

$$\tau = \frac{2\psi^3 - 3\psi^2 + 1}{6} + \epsilon\frac{1}{3}(1-\psi)^2 - \epsilon^2\frac{(1-\psi)^2}{180\psi^2} \tag{4.116}$$

Comparison with numerical solution. Equations (4.108)-(4.110), and (4.116) constitute the complete uniformly valid solution. Since the solution is implicit, the computation proceeds as follows. First, values of ϵ and r_f are fixed. Next, Eq. (4.110) is solved by iteration to obtain the value of ψ. Choosing values of ϕ in the range ψ to 1, Eqs. (4.109) and (4.108) are used to calculate the temperature distribution u as a function of r. The freezing time is calculated using Eq. (4.116).

Sample results for the temperature at the instant of complete freezing are shown in Fig. 4.2 for $\epsilon = 0.1$ and 0.5. For comparison, the corresponding numerical results of Tao (1967) are indicated by crosses. Even at $\epsilon = 0.1$, there exists some discrepancy between the perturbation and the numerical solutions. As discussed by Pedroso and Domoto (1973c), and Stephan and Holzknecht (1974), the error most probably lies in the numerical solution itself because for $\epsilon = 0.1$, one would expect the perturbation solution to be accurate.

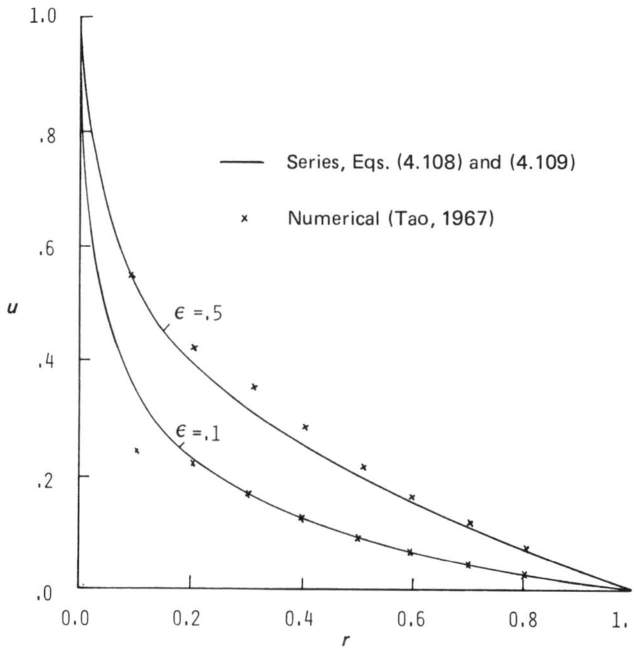

Figure 4.2 Temperature distribution in inward spherical solidification.

The results for freezing time appear in Fig. 4.3, and compare well with the numerical predictions of Tao (1967). It must be kept in mind that Tao's freezing time results are believed to be more accurate than his temperature results.

4.2 PRITULO'S METHOD

Since its original exposition in 1949, the spirit of Lighthill's technique has remained unchanged, but several approaches have been developed to simplify the procedure. One such simplification was introduced by Pritulo (1962). He proved that the straining expansion can be substituted directly into the ill-behaved perturbation expansion and straining functions then chosen in accordance with Lighthill's condition. In this way one solves algebraic rather than differential equations to determine the straining functions. We illustrate the procedure by reverting to the examples of the previous section.

4.2.1 A First-Order Differential Equation

Consider the perturbation expansion in Eq. (3.27) which is invalid in the vicinity of $x = 0$. The straining expansion is

$$x = s + \epsilon x_1 + \epsilon^2 x_2 \tag{4.117}$$

where x_1 and x_2 are functions of s.

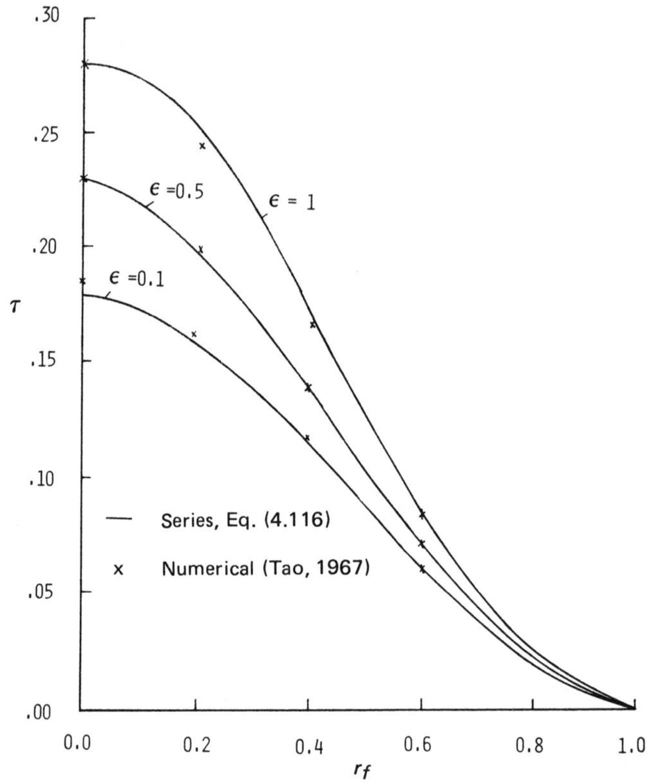

Figure 4.3 Strained coordinates and numerical solutions for freezing time in inward spherical solidification.

According to Pritulo's approach, we substitute Eq. (4.117) directly into Eq. (3.27). To do this, we need the expansion for powers of x. From Eq. (4.117)

$$x^n = s^n \left(1 + \epsilon \frac{x_1}{s} + \epsilon^2 \frac{x_2}{s} + \cdots\right)^n$$

Using binomial expansion and retaining terms up to $O(\epsilon^2)$, we get

$$x^n = s^n \left\{1 + \epsilon \frac{nx_1}{s} + \epsilon^2 \left[\frac{n(n-1)}{2!} \frac{x_1^2}{s^2} + \frac{nx_2}{s}\right]\right\} + O(\epsilon^3) \tag{4.118}$$

Making use of Eq. (4.118) in Eq. (3.27) and collecting terms with like powers of ϵ, the solution for y becomes

$$y = \frac{1}{s} + \epsilon \left(\frac{s^2-1}{2s^3} - \frac{x_2}{s^2}\right) - \epsilon^2 \left(\frac{s^2-1}{2s^5} + \frac{s^2-3}{2s^4} x_1 - \frac{x_1^2}{s^3} + \frac{x_2}{s}\right) + O(\epsilon^3) \tag{4.119}$$

To ensure that the singularity in the first-order term does not go beyond $1/s$, we put

$$-\frac{1}{2s^3} - \frac{x_1}{s^2} = 0 \qquad \text{or } x_1 = -\frac{1}{2s} \tag{4.120}$$

With this choice of x_1, the second-order term of Eq. (4.119) now reduces to $-\epsilon^2\,[(1/4s^3) + (x_2/s)]$. The only choice now is to eliminate this term completely giving

$$x_2 = -\frac{1}{4s^2} \tag{4.121}$$

Thus a uniformly valid solution is

$$y = \frac{1}{s} + \epsilon\,\frac{1}{2s} \tag{4.122}$$

$$x = s - \epsilon\,\frac{1}{2s} - \epsilon^2\,\frac{1}{4s^2} \tag{4.123}$$

which agrees with Eqs. (4.23) and (4.24) except for the second-order straining term in Eq. (4.123) which was not calculated previously.

As an alternative, one can eliminate the first-order term of Eq. (4.119) completely to give

$$x_1 = \frac{s^2 - 1}{2s} \tag{4.124}$$

From the second-order term of Eq. (4.119) it now follows that

$$x_2 = 0 \tag{4.125}$$

Thus

$$y = \frac{1}{s} \tag{4.126}$$

$$x = s + \epsilon\,\frac{s^2 - 1}{2s} \tag{4.127}$$

which agrees with Eqs. (4.31) and (4.32).

4.2.2 Melting of a Finite Slab

As another example, let us apply Pritulo's method to Eq. (3.77). After substituting Eq. (4.35) into Eq. (3.77), we need to expand terms like $e^{-n\tau}$. Using Eq. (4.35) it follows that

$$e^{-n\tau} = e^{-ns}\left[1 - \epsilon n\tau_1 + \epsilon^2\left(\frac{n^2\tau_1^2}{2} - n\tau_2\right)\right] + O(\epsilon^3) \tag{4.128}$$

Making use of Eq. (4.128) and grouping together terms with like powers of ϵ, the solution for v appears as

$$v = -\epsilon \tfrac{2}{3} e^{-3\tau} + \epsilon^2 [2\tau_1 e^{-3s} + \tfrac{8}{9}\mu(e^{-3s} - e^{-6s} - 3se^{-3s} + \tfrac{9}{2}s^2 e^{-3s})]$$

To remove the terms se^{-3s} and $s^2 e^{-3s}$, we put

$$2\tau_1 e^{-3s} - \tfrac{8}{3}\mu s e^{-3s} + 4\mu s^2 e^{-3s} = 0$$

or

$$\tau_1 = -\tfrac{2}{3}\mu s(3s-2) \tag{4.129}$$

which is in agreement with Eq. (4.61). With this choice, the solutions for v_2 and u_2 reduce to Eqs. (4.62) and (4.63), respectively.

4.2.3 Inward Spherical Solidification

From Eqs. (3.89) and (3.91), the two-term perturbation solution for u is

$$u = \frac{1-1/r}{1-1/r_f} + \epsilon\left[\frac{r_f^2 - 3r_f + 2}{6(1-r_f)^4}\left(1 - \frac{1}{r}\right) + \frac{r^2 - 3r + 2}{6r_f(1-r_f)^3}\right] \tag{4.130}$$

To render Eq. (4.130) uniformly valid, let us introduce the expansions, Eqs. (4.69) and (4.70), directly into Eq. (4.130) and retain terms up to $O(\epsilon)$. To this end, we need expansions for $1/r$ and $1/r_f$. These can be obtained using Eqs. (4.69) and (4.70) and carrying out the binomial expansions. Thus

$$\frac{1}{r} = \frac{1}{\phi} - \epsilon\frac{\sigma_1}{\phi^2} + O(\epsilon^2) \qquad \frac{1}{r_f} = \frac{1}{\psi} - \epsilon\frac{\hat{\sigma}_1}{\psi^2} + O(\epsilon^2)$$

Consider now the zero-order term in Eq. (4.130). It can be written as

$$\left(1 - \frac{1}{r}\right)\left(1 - \frac{1}{r_f}\right)^{-1} = \left(1 - \frac{1}{\phi} + \epsilon\frac{\sigma_1}{\phi^2}\right)\left(1 - \frac{1}{\psi} + \epsilon\frac{\hat{\sigma}_1}{\psi^2}\right)^{-1}$$

$$= \frac{1-1/\phi}{1-1/\psi} + \epsilon\left[\frac{\sigma_1}{\phi^2(1-1/\psi)} - \frac{\hat{\sigma}_1(1-1/\phi)}{\psi^2(1-1/\psi)^2}\right] + O(\epsilon^2)$$

For the first-order term in Eq. (4.130), we simply need to replace r by ϕ and r_f by ψ. Thus the two-term solution for u is

$$u = \frac{1-1/\phi}{1-1/\psi} + \epsilon\left[\frac{\sigma_1}{\phi^2(1-1/\psi)} - \frac{\hat{\sigma}_1(1-1/\phi)}{\psi^2(1-1/\psi)^2} + \frac{\psi^2 - 3\psi + 2}{6(1-\psi)^4}\left(1 - \frac{1}{\phi}\right) + \frac{\phi^2 - 3\phi + 2}{6\psi(1-\psi)^3}\right] + O(\epsilon^2) \tag{4.131}$$

To determine σ_1 we set the term in square brackets in Eq. (4.131) to zero giving

$$\frac{\sigma_1}{\phi^2(1-1/\psi)} - \frac{\hat{\sigma}_1(1-1/\phi)}{\psi^2(1-1/\psi)^2} + \frac{\psi^2-3\psi+2}{6(1-\psi)^4}\left(1-\frac{1}{\phi}\right) + \frac{\phi^2-3\phi+2}{6\psi(1-\psi)^3} = 0 \tag{4.132}$$

which can be rearranged as

$$\frac{\sigma_1}{\phi(\phi-1)} - \frac{\phi(\phi-2)}{6\psi^2(1-\psi)^2} = \frac{\psi^2-3\psi+2}{\psi(1-\psi)^3} - \hat{\sigma}_1\frac{1}{\psi(1-\psi)} \tag{4.133}$$

where it is noted that the right-hand side of Eq. (4.133) is a function of ψ alone.

Differentiating Eq. (4.133) with respect to ϕ, the differential equation for σ_1 is

$$\sigma_{1\phi} - \frac{2\phi-1}{\phi(\phi-1)}\sigma_1 = \frac{2\phi(\phi-1)^2}{6\psi^2(1-\psi)^2} \tag{4.134}$$

Integrating Eq. (4.134) we have

$$\frac{1}{\phi(\phi-1)}\sigma_1 = \frac{(\phi-1)^2}{6\psi^2(1-\psi)^2} + C$$

Imposing the condition $\lim_{\phi\to 1}(\sigma_1/\phi)_\phi = 0$ makes C vanish. Thus

$$\sigma_1 = \frac{-\phi(1-\phi)^3}{6\psi^2(1-\psi)^2} \tag{4.135}$$

which agrees with Eq. (4.104).

In an analogous manner, it is possible to derive the solution for σ_2 but the algebra becomes lengthy. The reader who is not overwhelmed by algebra can verify that the solution for σ_2 is in full agreement with Eq. (4.107).

4.3 MARTIN'S APPROACH

Pritulo's idea in the previous section was put in a new garb by Martin (1967). He exploited Lagrange's expansion to derive formulas which can be used directly with the original nonuniform expansion to give a uniformly valid solution. Here we derive one such formula and illustrate its use. Consider the expansions

$$y = y_0(x) + \epsilon y_1(x) + \epsilon^2 y_2(x) \tag{4.136}$$

$$x = s + \epsilon x_1(s) + \epsilon^2 x_2(s) \tag{4.137}$$

Substitute Eq. (4.137) into Eq. (4.136) and expand y_0, y_1, and y_2 into Taylor series about s. Retaining terms up to $O(\epsilon^2)$ we have

$$y = y_0(s) + [\epsilon x_1(s) + \epsilon^2 x_2(s)]y_0'(s) + \frac{\epsilon^2 x_1^2(s)}{2!} y_0''(s)$$
$$+ \epsilon[y_1(s) + \epsilon x_1(s)y_1'(s)] + \epsilon^2 y_2(s) + O(\epsilon^3)$$

Collecting terms of like powers of ϵ together, one obtains

$$y = y_0(s) + \epsilon[y_1(s) + x_1(s)y_0'(s)] + \epsilon^2[y_2(s) + x_1(s)y_1'(s)$$
$$+ x_2(s)y_0'(s) + \tfrac{1}{2}x_1^2(s)y_0''(s)] + O(\epsilon^3) \tag{4.138}$$

Let us choose $x_1(s)$ and $x_2(s)$ such that both the first-order and the second-order terms vanish. Thus

$$y_1(s) + x_1(s)y_0'(s) = 0 \tag{4.139}$$
$$y_2(s) + x_1(s)y_1'(s) + x_2(s)y_0'(s) + \tfrac{1}{2}x_1^2(s)y_0''(s) = 0 \tag{4.140}$$

which give

$$x_1(s) = -\frac{y_1(s)}{y_0'(s)} \tag{4.141}$$
$$x_2(s) = -\left\{\frac{y_2(s)}{y_0'(s)} + \frac{1}{2}\frac{y_1^2(s)y_0''(s)}{[y_0'(s)]^3} - \frac{y_1(s)y_1'(s)}{[y_0'(s)]^2}\right\} \tag{4.142}$$

Thus, a uniformly valid solution is

$$y = y_0(s) \tag{4.143}$$
$$x = s - \epsilon\frac{y_1(s)}{y_0'(s)} - \epsilon^2\left\{\frac{y_2(s)}{y_0'(s)} + \frac{1}{2}\frac{y_1^2(s)y_0''(s)}{[y_0'(s)]^3} - \frac{y_1(s)y_1'(s)}{[y_0'(s)]^2}\right\} + O(\epsilon^3) \tag{4.144}$$

where $y_n(s)$ are the functions appearing in the nonuniform expansions.

4.3.1 A First-Order Differential Equation

Let us render Eq. (3.27) uniformly valid using Eq. (4.144). The functions $y_n(s)$ are

$$y_0(s) = \frac{1}{s} \quad y_1(s) = \frac{s^2-1}{2s^3} \quad y_2(s) = -\frac{s^2-1}{2s^5} \tag{4.145}$$

and hence

$$y_0'(s) = -\frac{1}{s^2} \quad y_0''(s) = \frac{2}{s^3} \quad y_1'(s) = \frac{3-s^2}{2s^4} \tag{4.146}$$

Using Eqs. (4.145) and (4.146) in Eqs. (4.143) and (4.144) and simplifying, the final result is

$$y = \frac{1}{s} \tag{4.147}$$

$$x = s + \epsilon \frac{s^2 - 1}{2s} \tag{4.148}$$

Note that the second-order term in Eq. (4.144) vanishes. Equations (4.147) and (4.148) are in full agreement with Eqs. (4.31) and (4.32).

PROBLEMS

4.1 Consider Problem 3.2. Apply the method of strained coordinates to derive a uniformly valid solution of the form

$$y = s + \frac{b-1}{s} + \epsilon \frac{1}{2}\left[\frac{(b-1)^2}{s} - s\right]$$

$$x = s + \epsilon \frac{(1-b)}{2s}$$

Compare the foregoing with the exact solution

$$y = -\frac{x}{\epsilon} + \left[\frac{x^2}{\epsilon^2} + \frac{2}{\epsilon}(x^2 + b - 1) + b^2\right]^{1/2}$$

4.2 Use Lighthill's technique to show that a uniformly valid solution of

$$(x + \epsilon y)\frac{dy}{dx} + xy = be^{-x}$$

$$x = 1 \quad y = \frac{1}{e}$$

is

$$y = e^{-s}(b \ln s + 1) + O(\epsilon)$$

$$x = s - \epsilon(b \ln s + b + 1) + O(\epsilon^2)$$

4.3 Establish the result in Problem 4.1 by applying Pritulo's method to the nonuniform expansion of Problem 3.2.

4.4 Consider the nonuniform expansion given in Problem 3.4. Using Pritulo's method show that the simplest uniformly valid solution is

$$y = \frac{e^{-s}}{s^2}$$

$$x = s - \frac{\epsilon}{3s^2}$$

Note that

$$\int_1^x e^{-u}\left(\frac{2}{u^4}+\frac{1}{u^3}\right) du = \int_1^s e^{-u}\left(\frac{2}{u^4}+\frac{1}{u^3}\right) du$$

$$+ \int_s^{s+x_1(s)} e^{-u}\left(\frac{2}{u^4}+\frac{1}{u^3}\right) du$$

Hence show that at $x = 0$

$$y = \left(\frac{3}{\epsilon}\right)^{2/3} - \left(\frac{3}{\epsilon}\right)^{1/3} + O(1)$$

$$s = \left(\frac{\epsilon}{3}\right)^{1/3}$$

4.5 To render the expansion of Problem 3.5 valid in the region $r_f = 0$, strain the coordinates r and r_f and assume

$$u = u_0(\phi, \psi) + \epsilon u_1(\phi, \psi)$$

$$r = \phi + \epsilon\sigma_1(\phi, \psi)$$

$$r_f = \psi + \epsilon\sigma_1(\psi, \psi)$$

Following the analysis of Section 4.13 (see also Asfar et al., 1979)

$$u_0 = \frac{\ln\phi}{\ln\psi}$$

Making u_1 identically zero, reduce the first-order problem to

$$\frac{\partial}{\partial\phi}\left\{\phi\frac{\partial}{\partial\phi}[\sigma_1(\phi, \psi)]\right\} = \frac{\phi\ln\phi}{\psi^2\ln^2\psi}$$

and hence deduce the simplest solution for σ_1 as

$$\sigma_1(\phi, \psi) = \frac{\phi[(1-\phi^2) + (1+\phi^2)\ln\phi]}{4\psi^2\ln^2\psi}$$

Using the foregoing solution, show that

$$\tau = \frac{1}{4}(1-\psi^2) + \frac{1}{2}\psi^2\ln\psi + \epsilon\left[\frac{(1-\psi^2)+(1+\psi^2)\ln\psi}{2\ln\psi}\right] + O(\epsilon^2)$$

4.6 Reconsider the expansions given in Problem 3.5 and apply Pritulo's method to remove the singular behavior. The resulting solution should agree with the one appearing in Problem 4.5.

4.7 Use Lighthill's method as well as Pritulo's method to render the expansions for u in Problem 3.6 and τ in Problem 3.7 uniformly valid. To avoid excessive algebra, restrict your analysis to determining the first-order straining only. Both methods should, of course, lead to identical results.

CHAPTER

FIVE

METHOD OF MATCHED ASYMPTOTIC EXPANSIONS

5.1 INTRODUCTION

In this chapter, we concentrate on the application of the method of matched asymptotic expansion to the solution of those boundary value problems where the perturbation quantity ϵ appears in front of the term with the highest order of derivative. For such problems, a regular perturbation lowers the order of the differential equation which in turn means the solution cannot satisfy all the boundary conditions. A special treatment is therefore needed in the region near the boundary where its boundary condition is yet to be satisfied.

The origin of the method can be traced to the 1950s. Van Dyke (1975a) gives a brief summary of its development over the years and describes several applications to problems in fluid mechanics. A very detailed review of the various aspects of the method, from the point of view of an applied mathematician, has been given by Nayfeh (1973, 1981). Not all the variations of this method discussed in Nayfeh's book, however, have found application in heat transfer. We illustrate the method through several examples drawn from different areas of heat transfer.

5.2 HEAT TRANSFER FROM A MOVING ROD

We begin our discussion by treating a simple problem. Consider the heat transfer from a circular moving rod of radius R losing heat by convection as

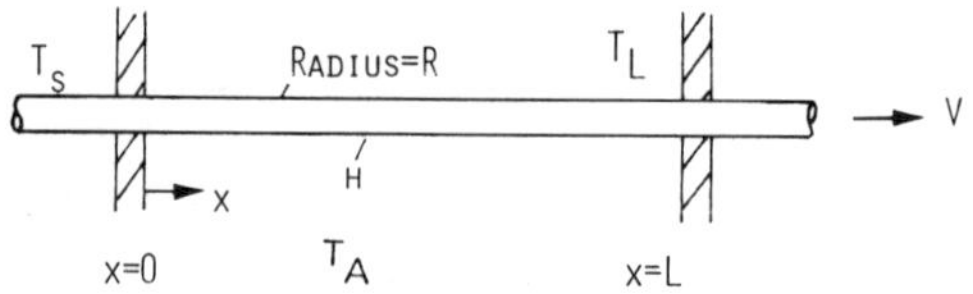

Figure 5.1 Heat transfer from a moving rod.

shown in Fig. 5.1. The rod moves at a constant speed V from one constant temperature bath at T_s to a lower temperature bath at T_L. As the rod moves, convection takes place to the environment at temperature T_a with heat transfer coefficient h. The energy equation, in dimensionless form, can be written as

$$\epsilon \frac{d^2\theta}{ds^2} + \frac{d\theta}{ds} - \beta\theta = 0 \tag{5.1}$$

subject to the boundary conditions

$$s = 0 \qquad \theta(0) = \theta_L \tag{5.2}$$

$$s = 1 \qquad \theta(1) = \theta_s \tag{5.3}$$

where

$$\epsilon = \frac{k}{\rho VLc} = \frac{1}{\text{Re}\,\text{Pr}} \qquad \beta = \frac{2hLT_a}{\rho VRc}$$
$$\theta_L = \frac{T_L - T_a}{T_a} \qquad \theta_s = \frac{T_s - T_a}{T_a} \tag{5.4}$$

and the dependent and independent variables are defined by

$$s = 1 - \frac{x}{L} \qquad \theta = \frac{T - T_a}{T_a} \tag{5.5}$$

where k = thermal conductivity
ρ = density
c = specific heat
Re = Reynolds number
Pr = Prandtl number

We now seek solutions of Eq. (5.1) for small values of ϵ. From the definition of ϵ, a small ϵ means either a large Reynolds number or a large Prandtl number or both. Since this small parameter appears in front of the term with the highest order of derivative, the order of the differential equation will be reduced by one as the parameter approaches zero. Let us introduce a regular perturbation series as

$$\theta(s) = \theta_0(s) + \epsilon\theta_1(s) + O(\epsilon^2) \tag{5.6}$$

Substituting Eq. (5.6) into Eq. (5.1) we get

$$\epsilon^0: \quad \frac{d\theta_0}{ds} - \beta\theta_0 = 0 \tag{5.7}$$

$$\epsilon^1: \quad \frac{d\theta_1}{ds} - \beta\theta_1 = -\frac{d^2\theta_0}{ds^2} \tag{5.8}$$

An inspection of Eqs. (5.7) and (5.8) shows that the perturbation equations are now one order-of-differentiation less than the original differential equation. Since a first-order differential equation requires one boundary condition to determine the integration constant, it is only necessary to satisfy one boundary condition. The question of which side ($s = 0$ or $s = 1$) should be chosen as the side the regular perturbation equations should satisfy is not an easy one. A systematic way is developed by Nayfeh (1973). Interested readers should follow through the examples in his book. In this problem, we choose the side $s = 1$, i.e., requiring the regular perturbation equations to satisfy the boundary condition in Eq. (5.3) and call the region where a regular perturbation is applicable the "outer" region.

Using Eq. (5.3) in Eq. (5.6) gives

$$\theta_0(1) = \theta_s \qquad \theta_1(1) = 0 \tag{5.9}$$

The solutions of Eqs. (5.7) and (5.8) are

$$\theta_0(s) = \theta_s e^{\beta(s-1)} \tag{5.10}$$

$$\theta_1(s) = -\beta^2\theta_s(s-1)e^{-\beta(1-s)} \tag{5.11}$$

The outer expansion is therefore

$$\theta(s) = \theta_s e^{\beta(s-1)} - \epsilon\beta^2\theta_s(s-1)e^{\beta(s-1)} + O(\epsilon^2)$$

or

$$\theta(s) = \theta_s e^{\beta(s-1)}[1 - \epsilon\beta^2(s-1)] + O(\epsilon^2) \tag{5.12}$$

Since we identify the "outer" region to be the region close to $s = 1$, we naturally identify the region close to $s = 0$ as the "inner" region. In fact, the outer region covers almost the whole region with the exception of a very narrow region close to $s = 0$.

To make the solution valid for the inner region, a transformation is needed to "move" the perturbation parameter ϵ away from the term of the highest derivative. If we introduce the transformation

$$\xi = \frac{s}{\epsilon} \qquad g(\xi) = \theta(s) \tag{5.13}$$

Equations (5.1) and (5.2) become

$$\frac{d^2g}{d\xi^2} + \frac{dg}{d\xi} = \epsilon\beta g \tag{5.14}$$

$$\xi = 0 \qquad g(0) = \theta_L \tag{5.15}$$

It is seen that the parameter ϵ has been moved away from the first term. A transformation such as the one defined by Eq. (5.13) in effect stretches the independent variable (or, the inner region). Let us now introduce the inner expansion as

$$g(\xi) = g_0(\xi) + \epsilon g_1(\xi) \tag{5.16}$$

Substituting the inner expansion in Eq. (5.16) into Eq. (5.14) we obtain

$$\epsilon^0: \quad \frac{d^2 g_0}{d\xi^2} + \frac{dg_0}{d\xi} = 0 \tag{5.17}$$

$$g_0(0) = \theta_L \tag{5.18}$$

$$\epsilon^1: \quad \frac{d^2 g_1}{d\xi^2} + \frac{dg_1}{d\xi} = \beta g_0 \tag{5.19}$$

$$g_1(0) = 0 \tag{5.20}$$

The solutions of Eqs. (5.17)-(5.20) are

$$g_0(\xi) = C_1(1 - e^{-\xi}) + \theta_L \tag{5.21}$$

$$g_1(\xi) = (C_1 + \theta_L)\beta\xi + C_2 - C_1\beta + [C_1\beta(\xi + 1) - C_2]e^{-\xi} \tag{5.22}$$

where C_1 and C_2 are constants which remain to be determined. The inner expansion is therefore

$$g(\xi) = [C_1(1 - e^{-\xi}) + \theta_L] + \epsilon\{(C_1 + \theta_L)\beta\xi + C_2 - C_1\beta + [C_1\beta(\xi + 1) - C_2]e^{-\xi}\} + O(\epsilon^2) \tag{5.23}$$

To determine C_1 and C_2, let us express the outer solution in terms of the inner variable ξ, Eq. (5.12) then becomes

$$\theta(s) = \theta_s e^{\beta(\epsilon\xi - 1)}[1 - \epsilon\beta^2(\epsilon\xi - 1)] + O(\epsilon^2) \tag{5.24}$$

which is then expanded for small ϵ while keeping ξ constant. The resulting solution, designated as $(\theta)^i$, is

$$(\theta)^i = \theta_s e^{-\beta}[1 + \beta\epsilon\xi + \epsilon\beta^2] + O(\epsilon^2) \tag{5.25}$$

Next, the inner solution, Eq. (5.23), is expressed in terms of the outer variable s to give

$$g = [C_1(1 - e^{-s/\epsilon}) + \theta_L] + \epsilon\left\{(C_1 + \theta_L)\beta\frac{s}{\epsilon} + C_2 - C_1\beta + \left[C_1\beta\left(\frac{s}{\epsilon} + 1\right) - C_2\right]e^{-s/\epsilon}\right\} + O(\epsilon^2)$$

or

$$g = [C_1(1 - e^{-s/\epsilon}) + \theta_L]$$
$$+ \{(C_1 + \theta_L)\beta s + \epsilon(C_2 - C_1\beta) + [C_1\beta s + \epsilon(C_1\beta - C_2)]e^{-s/\epsilon}\}$$
$$+ O(\epsilon^2) \tag{5.26}$$

which is then expanded for small ϵ while keeping s constant. The solution, designated as $(g)^o$, is

$$(g)^o = (C_1 + \theta_L) + [(C_1 + \theta_L)\beta s + \epsilon(C_2 - C_1\beta)] + O(\epsilon^2) \tag{5.27}$$

To match these two solutions we set

$$(\theta)^i = (g)^o \tag{5.28}$$

which is known as the matching principle. Equation (5.28) holds whenever two neighboring expansions have overlapping domains.

Substituting $(\theta)^i$ and $(g)^o$ from Eqs. (5.25) and (5.27), respectively, into Eq. (5.28), and reverting to the original variable s, we get

$$\theta_s e^{-\beta}(1 + \beta s + \epsilon\beta^2) = (C_1 + \theta_L) + (C_1 + \theta_L)\beta s + \epsilon(C_2 - C_1\beta) \tag{5.29}$$

which gives

$$\theta_s e^{-\beta} = C_1 + \theta_L$$
$$\theta_s e^{-\beta}\beta = (C_1 + \theta_L)\beta$$
$$\theta_s e^{-\beta}\beta^2 = C_2 - C_1\beta$$

from which C_1 and C_2 follow as

$$C_1 = \theta_s e^{-\beta} - \theta_L \tag{5.30}$$
$$C_2 = \theta_s e^{-\beta}(\beta + \beta^2) - \theta_L\beta \tag{5.31}$$

Since C_1 and C_2 are known, we can now write from Eqs. (5.12) and (5.23), respectively, the outer and inner expansions to order ϵ as

$$\theta(s) = \theta_s e^{\beta(s-1)}[1 - \epsilon\beta^2(s-1)] + O(\epsilon^2) \tag{5.32}$$

and

$$g(\xi) = [\theta_s e^{-\beta}(1 - e^{-\xi}) + \theta_L e^{-\xi}] + \epsilon\{\theta_s\beta e^{-\beta}(\xi + \beta)$$
$$+ [(\theta_s e^{-\beta} - \theta_L)\beta(\xi + 1) - (\beta\theta_s e^{-\beta} + \beta^2\theta_s e^{-\beta} - \theta_L\beta)]e^{-\xi}\} + O(\epsilon^2) \tag{5.33}$$

It should be noted that higher-order expansions can be obtained in a similar manner. Before closing the section, however, we will outline an alternative matching principle introduced by Van Dyke (1975a). Briefly, Van Dyke's matching principle states that

The m-term inner expansion of the n-term outer expansion

$$= \text{The } n\text{-term outer expansion of the } m\text{-term inner expansion} \quad (5.34)$$

For the example under consideration, $m = n = 2$, the steps to be followed are:

1. Two-term outer expansion
   From Eq. (5.12), we have

$$\theta(s) = \theta_s e^{\beta(s-1)}[1 - \epsilon\beta^2(s-1)]$$

   a. Rewrite in inner variable ξ

$$\theta(s) = \theta_s e^{\beta(\epsilon\xi - 1)}[1 - \epsilon\beta^2(\epsilon\xi - 1)]$$

   b. Expand for small ϵ (with ξ fixed)

$$\theta(s) = \theta_s e^{-\beta}(1 + \beta\epsilon\xi + \cdots)(1 + \epsilon\beta^2 - \epsilon^2\beta^2\xi)$$
$$= \theta_s e^{-\beta}[1 + \epsilon(\beta\xi + \beta^2) + \epsilon^2(\beta^3\xi - \beta^2\xi) - \epsilon^3\beta^3\xi^2]$$

   c. Two-term outer expansion

$$\theta(s) = \theta_s e^{-\beta}[1 + \epsilon\beta(\xi + \beta)] \quad (5.35)$$

2. Two-term inner expansion
   From Eq. (5.23), we have

$$g(\xi) = [C_1(1 - e^{-\xi}) + \theta_L] + \epsilon\{(C_1 + \theta_L)\beta\xi + C_2 - C_1\beta + [C_1\beta(\xi + 1) - C_2]e^{-\xi}\}$$

   a. Rewrite in outer variable s

$$g(\xi) = [C_1(1 - e^{-s/\epsilon}) + \theta_L] + \epsilon\left\{(C_1 + \theta_L)\beta\frac{s}{\epsilon} + C_2 - C_1\beta + \left[C_1\beta\left(\frac{s}{\epsilon} + 1\right) - C_2\right]e^{-s/\epsilon}\right\}$$

   b. Expand for small ϵ (with s fixed)

$$g(\xi) = (C_1 + \theta_L) + \epsilon\left[(C_1 + \theta_L)\beta\frac{s}{\epsilon} + C_2 - C_1\beta\right]$$

   c. Two-term inner expansion

$$g(\xi) = [(C_1 + \theta_L) + \beta s(C_1 + \theta_L)] + \epsilon[C_2 - C_1\beta] \quad (5.36)$$

Equating Eq. (5.35) to Eq. (5.36), the two constants C_1 and C_2 can be determined. The same results will be obtained as before.

To illustrate the order of magnitude of the inner and outer regions, let us compare the two solutions with the exact solution of Eq. (5.1), namely,

$$\theta(s) = C_3 e^{m_1 s} + C_4^{m_2 s} \tag{5.37}$$

where

$$\begin{aligned} m_1 &= \frac{1}{2\epsilon}(-1 + \sqrt{1 + 4\epsilon\beta}) \\ m_2 &= \frac{1}{2\epsilon}(-1 - \sqrt{1 + 4\epsilon\beta}) \\ C_3 &= \frac{\theta_s - \theta_L e^{m_2}}{e^{m_1} - e^{m_2}} \\ C_4 &= \theta_L - C_3 \end{aligned} \tag{5.38}$$

Table 5.1 gives a sample solution for $\theta_s = 1$, $\theta_L = 0$, $\beta = 0.01$, and $\epsilon = 0.001$. The outer solution is seen to deviate from the exact solution only in a small region close to $s = 0$.

5.3 HEAT CONDUCTION IN AN INSULATED CABLE

The heat conduction through an electrical cable with Joule heating is a nonlinear problem if the electrical and thermal conductivities are temperature dependent. Often the electrical conductor is surrounded by a thin layer of gas or an insulator whose surface temperature is kept constant. For low Joule heating, the problem can be solved by using the regular perturbation method without difficulty. For large Joule heating, however, the problem becomes singular and the method of matched asymptotic expansion provides a very efficient method for its solution.

From the work of Cohen and Shair (1970) the steady-state heat conduction equations for the cable and for the insulator can be written as

$$\frac{1}{r}\frac{d}{dr}\left[rk_0(1-\gamma T)\frac{dT}{dr}\right] + \sigma_0(1 - k_e T)E^2 = 0$$

$$\frac{1}{r}\frac{d}{dr}\left(r\frac{dT_i}{dr}\right) = 0$$

where E is the electrical field, $k_0(1-\gamma T)$ is the temperature-dependent thermal conductivity and $\sigma_0(1 - k_e T)$ is the electrical conductivity.

Table 5.1 Sample solution ($\theta_s = 1.0, \theta_L = 0.0, \beta = 0.01, \epsilon = 0.001$)

s	Exact	Inner	Outer
0.0000	0.0000	0.0000	0.9901
0.0002	0.1811	0.1795	0.9901
0.0004	0.3293	0.3264	0.9901
0.0006	0.4507	0.4467	0.9901
0.0008	0.5501	0.5452	0.9901
0.0010	0.6315	0.6258	0.9901
0.002	0.8638	0.8561	0.9901
0.004	0.9801	0.9720	0.9901
0.006	0.9965	0.9877	0.9901
0.008	0.9987	0.9898	0.9901
0.010	0.9990	0.9901	0.9901
0.02	0.9990	0.9902	0.9902
0.04	0.9990	0.9904	0.9904
0.06	0.9990	0.9906	0.9906
0.08	0.9991	0.9908	0.9908
0.10	0.9991	0.9910	0.9910
0.2	0.9992	0.9920	0.9920
0.4	0.9994	0.9940	0.9940
0.6	0.9996	0.9960	0.9960
0.8	0.9998	0.9980	0.9980
1.0	1.0000	1.0000	1.0000

By introducing the dimensionless variables

$$\eta = \frac{r}{r_c} \qquad \theta = \frac{T}{T_0} \qquad \theta_i = \frac{T_i}{T_0}$$

and

$$\alpha = \gamma T_0 \qquad \beta = k_e T_0 \qquad J = \frac{\sigma_0 E^2 r_c^2}{k_0 T_0}$$

the energy equations become

$$\frac{1}{\eta}\frac{d}{d\eta}\left[\eta(1-\alpha\theta)\frac{d\theta}{d\eta}\right] + (1-\beta\theta)J = 0 \qquad 0 \leqslant \eta \leqslant 1 \tag{5.39}$$

$$\frac{1}{\eta}\frac{d}{d\eta}\left(\eta\frac{d\theta_i}{d\eta}\right) = 0 \qquad 1 \leqslant \eta \leqslant \eta_0 \tag{5.40}$$

subject to the boundary conditions

$$\eta = 0 \qquad \frac{d\theta}{d\eta} = 0 \tag{5.41}$$

$$\eta = 1 \qquad \theta = \theta_i \tag{5.42}$$

$$\eta = 1 \qquad (1 - \alpha\theta)\frac{d\theta}{d\eta} = \gamma \frac{d\theta_i}{d\eta} \tag{5.43}$$

$$\eta = \eta_0 \qquad \theta_i = 0 \tag{5.44}$$

where J represents the rate of Joule heating.

The solution of Eq. (5.40) subject to the boundary condition, Eq. (5.44), is

$$\theta_i = C_0 \ln \frac{\eta}{\eta_0} \tag{5.45}$$

where C_0 is an unknown constant.

The boundary conditions, Eqs. (5.42) and (5.43), now become

$$\theta(1) = -C_0 \ln \eta_0 \tag{5.46}$$

$$[1 - a\theta(1)] \frac{d\theta(1)}{d\eta} = \gamma C_0 \tag{5.47}$$

or, using Eq. (5.46), the boundary condition in Eq. (5.47) becomes

$$\frac{d\theta(1)}{d\eta} + h\theta(1) = \alpha\theta(1)\frac{d\theta(1)}{d\eta} \tag{5.48}$$

where
$$h = \frac{\gamma}{\ln \eta_0}$$

The two boundary conditions therefore become

$$\frac{d\theta(1)}{d\eta} = \frac{\gamma C_0}{1 + \alpha C_0 \ln \eta_0} \tag{5.49}$$

$$\frac{d\theta(1)}{d\eta} + h\theta(1) = \alpha\theta(1)\frac{d\theta(1)}{d\eta} \tag{5.50}$$

The problem now becomes one of solving Eq. (5.39) and obtaining a constant C_0 subject to the boundary conditions in Eqs. (5.41), (5.49), and (5.50). For low Joule heating ($J \ll 1$) regular perturbation is appropriate. Such is not the case for high Joule heating ($J \gg 1$) since the problem now is singular. For $J \gg 1$, let us introduce ϵ as

$$\epsilon = \frac{1}{J} \tag{5.51}$$

Equation (5.39) then becomes

$$\epsilon \frac{d}{d\eta}\left[\eta(1 - \alpha\theta)\frac{d\theta}{d\eta}\right] + \eta(1 - \beta\theta) = 0 \tag{5.52}$$

which is clearly singular as $\epsilon \to 0$.

First, a regular perturbation is introduced as

$$\theta = \theta_0(\eta) + \epsilon\theta_1(\eta) + O(\epsilon^2) \tag{5.53}$$

Substituting Eq. (5.53) into Eq. (5.52) we get

$$\epsilon^0: \quad \beta\theta_0 - 1 = 0 \tag{5.54}$$

$$\epsilon^1: \quad \beta\eta\theta_1 = (1-\alpha\theta_0)\frac{d\theta_0}{d\eta} - \alpha\eta\frac{d\theta_0}{d\eta} + \eta(1-\alpha\theta_0)\frac{d^2\theta_0}{d\eta^2} \tag{5.55}$$

The solution of Eq. (5.54) is

$$\theta_0 = \frac{1}{\beta} \tag{5.56}$$

The solution of Eq. (5.55), as well as higher perturbations, is

$$\theta_1 = 0 \tag{5.57}$$

Close to the "boundary" region, a "stretched" independent variable is introduced:

$$\bar{\eta} = \frac{\eta - 1}{\epsilon^{1/2}} \tag{5.58}$$

Equation (5.52) becomes

$$\frac{d}{d\bar{\eta}}\left[(1+\epsilon^{1/2}\bar{\eta})(1-\alpha\theta)\frac{d\theta}{d\bar{\eta}}\right] + (1+\epsilon^{1/2}\bar{\eta})(1-\beta\theta) = 0 \tag{5.59}$$

and the boundary conditions become

$$\frac{d\theta(1)}{d\bar{\eta}} = \epsilon^{1/2}\left(\frac{\gamma C_0}{1+\alpha C_0 \ln \eta_0}\right) \tag{5.60}$$

$$[1-\alpha\theta(1)]\frac{d\theta(1)}{d\bar{\eta}} + \epsilon^{1/2}h\theta(1) = 0 \tag{5.61}$$

A perturbation expansion in terms of the "boundary" variable can be introduced as

$$\theta(\bar{\eta};\epsilon) = g_0(\bar{\eta}) + \epsilon^{1/2}g_1(\bar{\eta}) + \cdots \tag{5.62}$$

Equation (5.59) and the boundary condition in Eq. (5.60) then give

$$\epsilon^0: \quad \frac{d}{d\bar{\eta}}\left[(1-\alpha g_0)\frac{dg_0}{d\bar{\eta}}\right] + (1-\beta g_0) = 0 \tag{5.63}$$

$$\frac{dg_0(0)}{d\bar{\eta}} = 0 \tag{5.64}$$

$$^{1/2}: \quad \frac{d}{d\bar{\eta}}\left[(1-\alpha g_0)\frac{dg_1}{d\bar{\eta}} - \alpha\frac{dg_0}{d\bar{\eta}}g_1\right] - \beta g_1 = \frac{dg_0}{d\bar{\eta}} - \alpha g_0\frac{dg_0}{d\bar{\eta}} \tag{5.65}$$

$$\frac{dg_1(0)}{d\bar{\eta}} = \frac{\gamma C_0}{1+\alpha C_0 \ln \eta_0} \tag{5.66}$$

The solution of Eq. (5.63) is

$$g_0(\bar{\eta}) = \frac{1}{\beta} \tag{5.67}$$

In view of Eq. (5.67), Eq. (5.65) becomes

$$\frac{d}{d\bar{\eta}}\left(\frac{\beta-\alpha}{\beta}\frac{dg_1}{d\bar{\eta}}\right) - \beta g_1 = 0 \tag{5.68}$$

which has as its solution

$$g_1(\bar{\eta}) = C_1 \exp\left(\frac{\beta}{\sqrt{\beta-\alpha}}\,\bar{\eta}\right) + C_2 \exp\left(-\frac{\beta}{\sqrt{\beta-\alpha}}\,\bar{\eta}\right) \tag{5.69}$$

Applying the boundary condition, Eq. (5.66), we have

$$C_1 = C_2 + \frac{\gamma C_0\sqrt{\beta-\alpha}}{\beta(1+\alpha C_0 \ln \eta_0)}$$

and the solution for $g_1(\bar{\eta})$ takes the form

$$g_1(\bar{\eta}) = C_2\left[\exp\left(\frac{\beta}{\sqrt{\beta-\alpha}}\,\bar{\eta}\right) - \exp\left(-\frac{\beta}{\sqrt{\beta-\alpha}}\,\bar{\eta}\right)\right] + \frac{\gamma C_0\sqrt{\beta-\alpha}}{\beta(1+\alpha C_0 \ln \eta_0)}\exp\left(\frac{\beta}{\sqrt{\beta-\alpha}}\,\bar{\eta}\right) \tag{5.70}$$

In view of the solutions, Eqs. (5.56), (5.57), (5.67), and (5.70), the matching conditions are

$$\bar{\eta}\to-\infty: \qquad g_0(\bar{\eta}) \to \frac{1}{\beta} \tag{5.71}$$

$$g_1(\bar{\eta}) \to 0 \tag{5.72}$$

The matching condition, Eq. (5.71), is satisfied and the second matching condition, Eq. (5.72), can be satisfied if $C_2 = 0$. Equation (5.70) therefore reduces to

$$g_1(\bar{\eta}) = \frac{\gamma C_0\sqrt{\beta-\alpha}}{\beta(1+\alpha C_0 \ln \eta_0)}\exp\left(\frac{\beta}{\sqrt{\beta-\alpha}}\,\bar{\eta}\right) \tag{5.73}$$

From Eq. (5.62), the solution for θ follows as

$$\theta(\eta) = \frac{1}{\beta} + \epsilon^{1/2} \frac{\gamma C_0 \sqrt{\beta-\alpha}}{\beta(1+\alpha C_0 \ln \eta_0)} \exp\left[\frac{\epsilon^{1/2}\beta(\eta-1)}{\sqrt{\beta-\alpha}}\right] + O(\epsilon)$$

Finally, the constant C_0 can be determined by the boundary condition in Eq. (5.61) which gives

$$C_0 = -\frac{1}{\beta \ln \eta_0}\left[1 + \epsilon^{1/2}\frac{\gamma}{(\beta-\alpha)^{1/2}} + O(\epsilon)\right] \tag{5.74}$$

The asymptotic expansions of the temperature profiles for the conductor and for the insulator are, respectively,

For $0 \leqslant \eta \leqslant 1$

$$\theta(\eta) = \frac{1}{\beta} - \frac{1}{J^{1/2}} \frac{\gamma}{\beta(\beta-\alpha)^{1/2}\ln \eta_0} \exp\left\{\left[\frac{\beta J^{1/2}}{(\beta-\alpha)^{1/2}}\right](\eta-1)\right\} + O\left(\frac{1}{J}\right) \tag{5.75}$$

For $1 \leqslant \eta \leqslant \eta_0$

$$\theta_i(\eta) = \frac{1}{\beta \ln \eta_0}\left[1 + \frac{1}{J^{1/2}} \frac{\gamma}{(\beta-\alpha)^{1/2}\ln \eta_0} + O\left(\frac{1}{J}\right)\right] \ln \frac{\eta}{\eta_0} \tag{5.76}$$

5.4 FREEZING IN A FINITE SLAB

In this section, we will illustrate the application of the method of matched asymptotic expansion to a moving boundary problem. Specifically, let us consider the problem of inward freezing at large Stefan numbers of a finite slab initially at an arbitrary temperature above the freezing temperature, as shown in Fig. 5.2. The analysis follows that of Weinbaum and Jiji (1977).

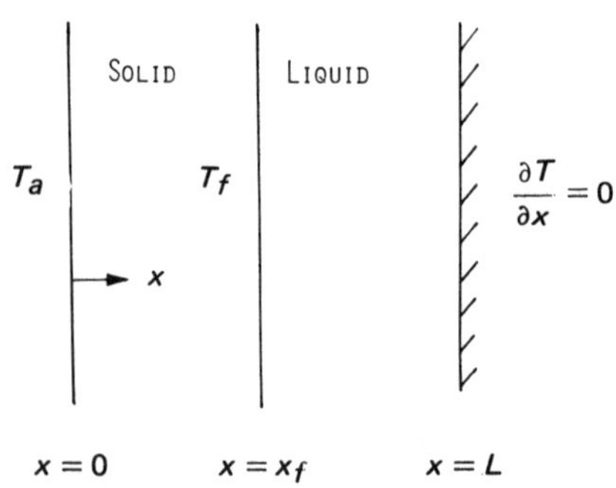

Figure 5.2 Freezing in a finite slab.

The energy equations of the solid and the liquid phases are

$$\alpha_s \frac{\partial^2 T_s}{\partial x^2} = \frac{\partial T_s}{\partial t} \tag{5.77}$$

$$\alpha_l \frac{\partial^2 T_l}{\partial x^2} = \frac{\partial T_l}{\partial t} \tag{5.78}$$

where subscripts s and l denote solid and liquid phases, respectively, and α is the thermal diffusivity. The boundary conditions for the two phases are

$$T_s(0, t) = T_a \tag{5.79}$$

$$T_s[x_f(t), t] = T_L(x_f(t), t] = T_f \tag{5.80}$$

$$\frac{\partial T_l(L, t)}{\partial x} = 0 \tag{5.81}$$

The initial conditions on the liquid temperature T_l and interface location $x_f(t)$ are

$$T_l(x, 0) = T_i \tag{5.82}$$

$$x_f(0) = 0 \tag{5.83}$$

The energy balance at the interface requires

$$k_s \frac{\partial T_s(x_f, t)}{\partial x} - k_l \frac{\partial T_l(x_f, t)}{\partial x} = \rho H \frac{dx_f}{dt} \tag{5.84}$$

By introducing the dimensionless quantities

$$\theta = \frac{k_s(T_s - T_f)}{k_l(T_i - T_f)} \qquad \phi = \frac{T_l - T_f}{T_i - T_f} \qquad \xi = \frac{x}{x_f(t)}$$
$$\eta = \frac{x - x_f}{L - x_f} \qquad \tau = \frac{\epsilon \alpha_l t}{L^2} \qquad \sigma = \frac{x_f}{L} \qquad \epsilon = \frac{c(T_i - T_f)}{H} \tag{5.85}$$

Equations (5.77)-(5.83) become

$$\frac{\partial^2 \theta}{\partial \xi^2} = \epsilon \left(\frac{\alpha_l}{\alpha_s}\right) \left(\sigma^2 \frac{\partial \theta}{\partial \tau} - \xi \sigma \frac{d\sigma}{d\tau} \frac{\partial \theta}{\partial \xi}\right) \tag{5.86}$$

$$\frac{\partial^2 \phi}{\partial \eta^2} = \epsilon \left[(1 - \sigma)^2 \frac{\partial \phi}{\partial \tau} - (1 - \sigma)(1 - \eta) \frac{d\sigma}{d\tau} \frac{\partial \phi}{\partial \eta}\right] \tag{5.87}$$

subject to the boundary conditions

$$\theta(0, \tau) = \frac{k_s(T_a - T_f)}{k_l(T_i - T_f)} = \theta_a \tag{5.88}$$

$$\theta(1, \tau) = \phi(0, \tau) = 0 \tag{5.89}$$

$$\frac{\partial \phi(1,\tau)}{\partial \eta} = 0 \tag{5.90}$$

and the initial conditions

$$\phi(\eta, 0) = 1 \tag{5.91}$$

$$\sigma(0) = 0 \tag{5.92}$$

The energy balance at the interface, Eq. (5.84), becomes

$$(1-\sigma)\frac{\partial \theta(1,\tau)}{\partial \xi} - \sigma\frac{\partial \phi(0,\tau)}{\partial \eta} = \sigma(1-\sigma)\frac{d\sigma}{d\tau} \tag{5.93}$$

Mathematically, we seek solutions to Eqs. (5.86) and (5.87), subject to the boundary and initial conditions, Eqs. (5.88)-(5.92), and the interfacial energy balance, Eq. (5.93), for a given value of θ_a. The ratio of the diffusivities α_l/α_s is, of course, known. The perturbation parameter ϵ is assumed to be small. Since the parameter ϵ appears in front of the highest order of derivatives of θ_0 and ϕ with respect to τ, the initial conditions, Eqs. (5.91) and (5.92), cannot be satisfied. An inner expansion near $\tau = 0$ will have to be introduced in addition to the regular perturbation.

Let us proceed with an outer expansion of the form

$$\theta(\xi, \tau; \epsilon) = \theta_0(\xi, \tau) + \epsilon^{1/2}\theta_1(\xi, \tau) + O(\epsilon) \tag{5.94}$$

$$\phi(\eta, \tau; \epsilon) = \phi_0(\xi, \tau) + \epsilon^{1/2}\phi_1(\xi, \tau) + O(\epsilon) \tag{5.95}$$

$$\sigma(\tau; \epsilon) = \sigma_0(\tau) + \epsilon^{1/2}\sigma_1(\tau) + O(\epsilon) \tag{5.96}$$

which, upon substitution into Eqs. (5.86), (5.87), and (5.93), gives

$$\epsilon^0: \quad \frac{\partial^2 \theta_0}{\partial \xi^2} = 0 \tag{5.97}$$

$$\frac{\partial^2 \phi_0}{\partial \eta^2} = 0 \tag{5.98}$$

$$(1-\sigma_0)\frac{\partial \theta_0(1,\tau)}{\partial \xi} - \sigma_0\frac{\partial \phi_0(0,\tau)}{\partial \eta} = \sigma_0(1-\sigma_0)\frac{d\sigma_0}{d\tau} \tag{5.99}$$

$$\epsilon^{1/2}: \quad \frac{\partial^2 \theta_1}{\partial \xi^2} = 0 \tag{5.100}$$

$$\frac{\partial^2 \phi_1}{\partial \eta^2} = 0 \tag{5.101}$$

$$(1-\sigma_0)\frac{\partial \theta_1(1,\tau)}{\partial \xi} - \sigma_1\frac{\partial \theta_0(1,\tau)}{\partial \xi} - \sigma_0\frac{\partial \phi_1(0,\tau)}{\partial \eta} - \sigma_1\frac{\partial \phi_0(0,\tau)}{\partial \eta}$$

$$= \sigma_0(1-\sigma_0)\frac{d\sigma_1}{d\tau} - \sigma_0\sigma_1\frac{d\sigma_0}{d\tau} + \sigma_1(1-\sigma_0)\frac{d\sigma_0}{d\tau} \tag{5.102}$$

The solutions of Eqs. (5.97)-(5.102) can be obtained easily as

$$\theta_0 = \theta_a(1 - \xi) \tag{5.103}$$

$$\theta_1 = 0 \tag{5.104}$$

$$\phi_0 = 0 \tag{5.105}$$

$$\phi_1 = 0 \tag{5.106}$$

$$\sigma_0 = (C_0 - 2\theta_a\tau)^{1/2} \tag{5.107}$$

$$\sigma_1 = 0 \tag{5.108}$$

The outer expansions are therefore

$$\theta(\xi, \tau; \epsilon) = \theta_a(1 - \xi) + O(\epsilon) \tag{5.109}$$

$$\phi(\eta, \tau; \epsilon) = 0 + O(\epsilon) \tag{5.110}$$

$$\sigma(\tau; \epsilon) = (C_0 - 2\theta_a\tau)^{1/2} + O(\epsilon) \tag{5.111}$$

For the inner expansions, the following inner variables are introduced

$$\begin{aligned} T &= \frac{\tau}{\epsilon} \\ \hat{\theta}(\xi, T) &= \theta(\xi, \tau) \\ \hat{\phi}(\eta, T) &= \phi(\eta, \tau) \\ \hat{\sigma}(T) &= \sigma(\tau) \end{aligned} \tag{5.112}$$

into Eqs. (5.86), (5.87), and (5.93) to give

$$\frac{\partial^2\hat{\theta}}{\partial\xi^2} = \frac{\alpha_l}{\alpha_s}\left(\hat{\sigma}^2\,\frac{\partial\hat{\theta}}{\partial T} - \xi\hat{\sigma}\,\frac{d\hat{\sigma}}{dT}\,\frac{\partial\hat{\theta}}{\partial\xi}\right) \tag{5.113}$$

$$\frac{\partial^2\hat{\phi}}{\partial\eta^2} = (1 - \hat{\sigma})^2\,\frac{\partial\hat{\phi}}{\partial T} - (1 - \hat{\sigma})(1 - \eta)\,\frac{d\hat{\sigma}}{dT}\,\frac{\partial\hat{\phi}}{\partial\eta} \tag{5.114}$$

$$\epsilon\left[(1 - \hat{\sigma})\,\frac{\partial\hat{\theta}(1, T)}{\partial\xi} - \hat{\sigma}\,\frac{\partial\hat{\phi}(0, T)}{\partial\eta}\right] = \hat{\sigma}(1 - \hat{\sigma})\,\frac{d\hat{\sigma}}{dT} \tag{5.115}$$

By expanding $\hat{\theta}$, $\hat{\phi}$, and $\hat{\sigma}$ in terms of inner variables,

$$\hat{\theta}(\xi, T; \epsilon) = \hat{\theta}_0(\xi, T) + \epsilon^{1/2}\hat{\theta}_1(\xi, T) + O(\epsilon) \tag{5.116}$$

$$\hat{\phi}(\eta, T; \epsilon) = \hat{\phi}_0(\eta, T) + \epsilon^{1/2}\hat{\phi}_1(\eta, T) + O(\epsilon) \tag{5.117}$$

$$\hat{\sigma}(T) = \hat{\sigma}_0(T) + \epsilon^{1/2}\hat{\sigma}_1(T) + O(\epsilon) \tag{5.118}$$

Equations (5.113)-(5.115) lead to

$$\epsilon^0: \quad \frac{\partial^2 \hat{\theta}_0}{\partial \xi^2} = 0 \tag{5.119}$$

$$\frac{\partial^2 \hat{\phi}_0}{\partial \eta^2} = \frac{\partial \phi_0}{\partial T} \tag{5.120}$$

$$\hat{\sigma}_0 = 0 \tag{5.121}$$

$$\epsilon^{1/2}: \quad \frac{\partial^2 \hat{\theta}_1}{\partial \xi^2} = 0 \tag{5.122}$$

$$\frac{\partial^2 \hat{\phi}_1}{\partial \eta^2} = \frac{\partial \hat{\phi}_1}{\partial T} - 2\hat{\sigma}_1 \frac{\partial \hat{\phi}_0}{\partial T} - (1-\eta)\frac{d\hat{\sigma}_1}{dT}\frac{\partial \hat{\phi}_0}{\partial \eta} \tag{5.123}$$

$$\frac{\partial \hat{\theta}_0(1,T)}{\partial \xi} = \hat{\sigma}_1 \frac{d\hat{\sigma}_1}{dT} \tag{5.124}$$

The solutions of Eqs. (5.119)-(5.124) are

$$\hat{\theta}_0 = \theta_a(1-\xi) \tag{5.125}$$

$$\hat{\theta}_1 = 0 \tag{5.126}$$

$$\hat{\phi}_0 = \sum_{j=0}^{\infty} \left[\frac{2}{(j+\frac{1}{2})\pi} \sin\left(j+\frac{1}{2}\right)\pi\eta\right] \exp\left\{-\left[\left(j+\frac{1}{2}\right)\pi\right]^2 T\right\} \tag{5.127}$$

$$\hat{\sigma}_0 = 0 \tag{5.128}$$

$$\hat{\sigma}_1 = (-2\theta_a T)^{1/2} \tag{5.129}$$

Here, the solution of $\hat{\phi}_1$ is not presented due to the fact that it is very complicated. Furthermore, it is not needed in the solution of σ_2 which is considered next. The equation governing σ_2 is

$$\frac{\partial \hat{\theta}_1(1,T)}{\partial \xi} - \hat{\sigma}_1 \frac{\partial \hat{\theta}_0(1,T)}{\partial \xi} - \hat{\sigma}_1 \frac{\partial \hat{\phi}_0(0,T)}{\partial \eta} = \frac{d}{dT}(\hat{\sigma}_1\hat{\sigma}_2) - \hat{\sigma}_1^2 \frac{d\hat{\sigma}_1}{dT} \tag{5.130}$$

The solution of Eq. (5.130) is

$$\hat{\sigma}_2 = \sum_{j=0}^{\infty} \left(\frac{2}{[(j+\frac{1}{2})\pi]^2} \exp\left\{-\left[\left(j+\frac{1}{2}\right)\pi\right]^2 T\right\} - \frac{\sqrt{\pi}\,\mathrm{erf}\,[(j+\frac{1}{2})\pi\sqrt{T}]}{[(j+\frac{1}{2})\pi]^3\sqrt{T}}\right) \tag{5.131}$$

It is therefore seen that θ and $\hat{\theta}$ are identical to $O(\epsilon)$. No matching is needed. To match the outer and inner solutions of the dimensionless interface location, σ and $\hat{\sigma}$, the matching condition

$$\sigma^{(i)} = \hat{\sigma}^{(o)}$$

can be used and we get $C_0 = 0$.

We therefore have the outer solutions to order ϵ as

$$\theta(\xi, \tau; \epsilon) = \theta_a(1-\xi) \tag{5.132}$$

$$\sigma(\tau; \epsilon) = (-2\theta_a \tau)^{1/2} \tag{5.133}$$

and the inner solutions to order $\epsilon^{3/2}$

$$\hat{\theta}(\xi, T; \epsilon) = \theta_a(1-\xi) \tag{5.134}$$

$$\hat{\sigma}(T) = (-2\theta_a \tau)^{1/2} + \sum_{j=0}^{\infty} \left(\frac{2}{[(j+\frac{1}{2})\pi]^2} \exp\left\{ -\left[\left(j+\frac{1}{2}\right)\pi\right]^2 \frac{\tau}{\epsilon} \right\} - \frac{\sqrt{\pi}\,\text{erf}\,[(j+\frac{1}{2})\pi\sqrt{\tau/\epsilon}]}{[(j+\frac{1}{2})\pi]^3 \sqrt{\tau/\epsilon}} \right) \tag{5.135}$$

For physical interpretation of the solutions, the reader is referred to Weinbaum and Jiji (1977).

5.5 INTERACTION OF RADIATION WITH NATURAL CONVECTION

We will next consider the problem by R. D. Cess (1966) involving the convective phenomena of fluids which absorb and emit thermal radiation. Assuming the plate surface to be black and the fluid to be gray and nonscattering, the boundary-layer equation of the natural convection flow over a vertical semi-infinite flat plate can be written in terms of the stream function, ψ, defined by

$$u = \frac{\partial \psi}{\partial y} \qquad v = -\frac{\partial \psi}{\partial x}$$

as

$$\frac{\partial \psi}{\partial y}\frac{\partial^2 \psi}{\partial x \partial y} - \frac{\partial \psi}{\partial x}\frac{\partial^2 \psi}{\partial y^2} = \nu \frac{\partial^3 \psi}{\partial y^3} + g\beta(T - T_\infty)$$

$$\frac{\partial \psi}{\partial y}\frac{\partial T}{\partial x} - \frac{\partial \psi}{\partial x}\frac{\partial T}{\partial y} = \alpha \frac{\partial^2 T}{\partial y^2} - \frac{1}{\rho C_p}\frac{\partial q_R}{\partial y}$$

where q_R denotes the radiation flux within the fluid in the y direction and is given by

$$-\frac{\partial q_R}{\partial \tau} = 2\sigma T_\infty^4 E_2(\tau) + 2\sigma \int_0^\infty T^4(x,t) E_1(|\tau - t|)\,dt - 4\sigma T^4(x,\tau)$$

Introducing the dimensionless variables

$$\xi = \frac{4\sigma\kappa T_\infty^3 x^{1/2}}{\rho C_p (g\beta\Delta T)^{1/2}} \qquad N = \frac{k\kappa}{4\sigma T_\infty^3}$$

$$\psi(x,y) = \left(\frac{g\beta\Delta T x}{\kappa}\right)^{1/2} f(\epsilon,\tau)$$

$$T(x,y) = T_\infty \theta(\xi,\tau)$$

the boundary-layer equations become

$$\left(\frac{\partial f}{\partial \tau}\right)^2 + \xi\frac{\partial f}{\partial \tau}\frac{\partial^2 f}{\partial \xi\,\partial \tau} - f\frac{\partial^2 f}{\partial \tau^2} - \xi\frac{\partial f}{\partial \xi}\frac{\partial^2 f}{\partial \tau^2} = 2\,\mathrm{Pr}\,N\xi\frac{\partial^3 f}{\partial \tau^3} + 2\frac{\theta - 1}{\theta_w - 1} \tag{5.136}$$

$$\frac{\partial f}{\partial \tau}\frac{\partial \theta}{\partial \xi} - \frac{1}{\xi}f + \frac{\partial f}{\partial \xi}\frac{\partial \theta}{\partial \tau} = 2N\frac{\partial^2 \theta}{\partial \tau^2}$$

$$+\,\theta_w^4 E_2(\tau) + \int_0^\infty \theta^4(\xi,t) E_1(|\tau - t|)\,dt - 2\theta^4(\xi,\tau) \tag{5.137}$$

subject to the boundary conditions

$$\tau = 0 \qquad f(\xi,0) = \frac{\partial f(\xi,0)}{\partial \tau} = 0 \qquad \theta(\xi,0) = \theta_w \tag{5.138}$$

$$\tau = \infty \qquad \frac{\partial f(\xi,\infty)}{\partial \tau} = 0 \qquad \theta(\xi,\infty) = 1 \tag{5.139}$$

where $E_n(t)$ is the exponential integral defined by

$$E_n(t) = \int_0^1 \mu^{n-2} e^{-t/\mu}\,d\mu$$

and the dimensionless parameter N represents a measure of the importance of conduction versus radiation within the fluid. Cess estimated that, for water vapor, carbon dioxide, and ammonia, where N has been calculated using the Planck mean absorption coefficient the magnitude of N is found to be much less than unity, i.e., $N \ll 1$. Since N appears in front of terms of the highest order of derivatives, the boundary value problem is singular and the method of matched asymptotic expansion is needed.

To get the outer expansion, a regular perturbation of f and θ with N as the perturbation parameter is usually made. However, due to the complexity of the problem, Cess took only the first term of the expansion. The equation for the solution of these first terms of f and θ, represented by F and H, can be obtained by setting $N = 0$ in Eqs. (5.136) and (5.137), which gives

$$\left(\frac{\partial F}{\partial \tau}\right)^2 + \xi \frac{\partial F}{\partial \tau}\frac{\partial^2 F}{\partial \xi \, \partial \tau} - F\frac{\partial^2 F}{\partial \tau^2} - \xi \frac{\partial F}{\partial \xi}\frac{\partial^2 F}{\partial \tau^2} = 2\left(\frac{H-1}{\theta_w - 1}\right) \tag{5.140}$$

$$\frac{\partial F}{\partial \tau}\frac{\partial H}{\partial \xi} - \left(\frac{F}{\xi} + \frac{\partial F}{\partial \xi}\right)\frac{\partial H}{\partial \tau} = \theta_w^4 E_2(\tau) + \int_0^\infty H^4(\xi, t)E_1(|\tau - t|)\,dt - 2H^4(\xi, \tau) \tag{5.141}$$

The boundary conditions to be satisfied are the "outer" conditions

$$\tau = \infty \qquad \frac{\partial F(\xi, \infty)}{\partial \tau} = 0 \qquad H(\xi, \infty) = 1 \tag{5.142}$$

For the inner layer, we again "stretch" the variables by introducing

$$\bar{\tau} = \frac{\tau}{\sqrt{2 \Pr N}} \qquad \bar{f} = \frac{f}{\sqrt{2 \Pr N}} \tag{5.143}$$

Equations (5.136) and (5.137) become

$$\left(\frac{\partial \bar{f}}{\partial \bar{\tau}}\right)^2 + \xi \frac{\partial \bar{f}}{\partial \bar{\tau}}\frac{\partial^2 \bar{f}}{\partial \xi \, \partial \bar{\tau}} - \bar{f}\frac{\partial^2 \bar{f}}{\partial \bar{\tau}^2} - \xi \frac{\partial \bar{f}}{\partial \xi}\frac{\partial^2 \bar{f}}{\partial \bar{\tau}^2} = \xi \frac{\partial^3 \bar{f}}{\partial \bar{\tau}^3} + 2\frac{\theta - 1}{\theta_w - 1} \tag{5.144}$$

$$\frac{\partial \bar{f}}{\partial \bar{\tau}}\frac{\partial \theta}{\partial \xi} - \left(\frac{\bar{f}}{\xi} + \frac{\partial \bar{f}}{\partial \xi}\right)\frac{\partial \theta}{\partial \bar{\tau}} = \frac{1}{\Pr}\frac{\partial^2 \theta}{\partial \bar{\tau}^2} + \theta_w^4 + \int_0^\infty H^4(\xi, t)E_1(t)\,dt - 2\theta^4(\xi, \tau) \tag{5.145}$$

Equation (5.145) represents a simplified form of Eq. (5.137) for optically thin radiation, which is the case in the inner region.

The inner boundary conditions are

$$\bar{\tau} = 0 \qquad \bar{f}(\xi, 0) = 0 \qquad \frac{\partial \bar{f}(\xi, 0)}{\partial \bar{\tau}} = 0 \qquad \theta(\xi, 0) = \theta_w \tag{5.146}$$

The solution of the outer and inner problems is not an easy task. For this reason, Cess restricted his solutions to small ξ, i.e., second-order interactions of radiation with convection. For the outer solution, let us expand the variables so that

$$F(\xi, \tau) = \Gamma \xi^{1/3}\left[F_0(\tau) + F_1(\tau)\xi^{2/3} + \cdots\right] \tag{5.147}$$

$$H(\xi, \tau) = 1 + (\theta_w - 1)\Gamma^2 \xi^{2/3}\left[G_0(\tau) + G_1(\tau)\xi^{2/3} + \cdots\right] \tag{5.148}$$

where

$$\Gamma = \left(\frac{\theta_w^4 - 1}{\theta_w - 1} \right)^{1/3} \tag{5.149}$$

By substituting the expansions, Eqs. (5.147) and (5.148), into Eqs. (5.140) and (5.141), and separating terms with different powers of $\xi^{2/3}$, we get

$$\xi^{2/3}: \quad F_0 F_0'' - (F_0')^2 + 2G_0 = 0 \tag{5.150}$$

$$2F_0 G_0' - F_0' G_0 + \frac{3}{2} E_2(\tau) = 0 \tag{5.151}$$

$$\xi^{4/3}: \quad F_0 F_1'' - \frac{5}{2} F_0' F_1' + \frac{3}{2} F_0'' F_1 + \frac{3}{2} G_1 = 0 \tag{5.152}$$

$$F_0 G_1' - F_0' G_1 = \frac{1}{2} F_1' G_0 - \frac{3}{2} F_1 G_0' + \frac{3}{\Gamma} \left[2G_0 - \int_0^\infty G_0(t) E_1(|\tau - t|)\, dt \right] \tag{5.153}$$

where primes denote differentiation with respect to τ. The outer boundary conditions are

$$\tau = \infty \qquad G_0(\infty) = G_1(\infty) = F_0'(\infty) = F_1'(\infty) = 0 \tag{5.154}$$

For the inner region, $\bar{f}$ and θ can be expanded so that

$$\bar{f}(\xi, \bar{\tau}) = 2\xi^{1/2} [f_0(\eta) + f_1(\eta)\Gamma\xi^{1/3} + f_2(\eta)\Gamma^2\xi^{2/3} + \cdots] \tag{5.155}$$

$$\theta(\xi, \bar{\tau}) = 1 + (\theta_w - 1)[\theta_0(\eta) + \theta_1(\eta)\Gamma\xi^{1/3} + \theta_2(\eta)\Gamma^2\xi^{2/3} + \cdots] \tag{5.156}$$

where

$$\eta = \frac{\bar{\tau}}{\xi^{1/2}} \tag{5.157}$$

By substituting the expansions, Eqs. (5.155) and (5.156), into Eqs. (5.144) and (5.145), and separating the various orders of perturbation, we get

$$\xi^0: \quad f_0''' + 3f_0 f_0'' - 2(f_0')^2 + \theta_0 = 0 \tag{5.158}$$

$$\frac{1}{\mathrm{Pr}} \theta_0'' + 3f_0 \theta_0' = 0 \tag{5.159}$$

$$\xi^{1/3}: \quad f_1''' + 3f_0 f_1'' - \frac{14}{3} f_0' f_1' + \frac{11}{3} f_0'' f_1 + \theta_1 = 0 \tag{5.160}$$

$$\frac{1}{\mathrm{Pr}} \theta_1'' + 3f_0 \theta_1' - \frac{2}{3} f_0' \theta_1 + \frac{11}{3} f_1 \theta_0' = 0 \tag{5.161}$$

$$\xi^{2/3}: \quad f_2''' + 3f_0 f_2'' - \frac{16}{3} f_0' f_2' + \frac{13}{3} f_0'' f_2 = -\theta_2 + \frac{8}{3}(f_1')^2 - \frac{11}{3} f_1 f_1'' \tag{5.162}$$

$$\frac{1}{\Pr}\theta_2'' + 3f_0\theta_2' - \frac{4}{3}f_0'\theta_2 = -\frac{11}{3}f_1\theta_1' - \frac{13}{3}f_2\theta_0' + \frac{2}{3}f_1'\theta_1 \tag{5.163}$$

where primes denote differentiation with respect to η. The inner boundary conditions are

$$f_0(0) = f_1(0) = f_2(0) = 0 \tag{5.164}$$

$$f_0'(0) = f_1'(0) = f_2'(0) = 0 \tag{5.165}$$

$$\theta_0(0) = 1 \qquad \theta_1(0) = \theta_2(0) = 0 \tag{5.166}$$

To get the inner boundary conditions of the outer solution and the outer boundary conditions of the inner solution, the matching conditions are used. They are

$$F(\xi, 0) = 0 \tag{5.167}$$

$$\frac{\partial \bar{f}(\xi, \infty)}{\partial \bar{\tau}} = \frac{\partial F(\xi, 0)}{\partial \tau} \tag{5.168}$$

$$\theta(\xi, \infty) = H(\xi, 0) \tag{5.169}$$

Substituting $F, \bar{f}, \theta$, and H from Eqs. (5.147), (5.155), (5.148), and (5.156) into the matching conditions, Eqs. (5.167)-(5.169), we get

$$F_0(0) + F_1(0)\xi^{2/3} + \cdots = 0 \tag{5.170}$$

$$\begin{aligned} 2[f_0'(\infty) + f_1'(\infty)\Gamma\xi^{1/3} + f_2'(\infty)\Gamma^2\xi^{2/3} + \cdots] \\ = \Gamma\xi^{1/3}[F_0'(0) + F_1'(0)\xi^{2/3} + \cdots] \end{aligned} \tag{5.171}$$

$$\begin{aligned} 1 + (\theta_w - 1)[\theta_0(\infty) + \theta_1(\infty)\Gamma\xi^{1/3} + \theta_2(\infty)\Gamma^2\xi^{2/3} + \cdots] \\ = 1 + (\theta_w - 1)\Gamma^2\xi^{2/3}[G_0(0) + G_1(0)\xi^{2/3} + \cdots] \end{aligned} \tag{5.172}$$

Equating terms of like powers of ξ and Γ, we get the inner boundary conditions for the outer expansion

$$F_0(0) = F_1(0) = 0 \tag{5.173}$$

from Eq. (5.170) and the outer boundary conditions for the inner solution

$$f_0'(\infty) = 0 \qquad f_1'(\infty) = \tfrac{1}{2}F_0'(0) \qquad f_2'(\infty) = 0 \tag{5.174}$$

$$\theta_0(\infty) = \theta_1(\infty) = 0 \qquad \theta_2(\infty) = G_0(0) \tag{5.175}$$

from Eqs. (5.171) and (5.172), respectively.

The solution procedure of this problem is as follows:

1. Solve the outer equations, Eqs. (5.150)-(5.153), subject to the boundary conditions, Eqs. (5.154) and (5.173), independent of the inner solution.
2. Solve the inner equations, Eqs. (5.158)-(5.163), subject to the boundary conditions, Eqs. (5.164)-(5.166) and Eqs. (5.174) and (5.175). The solutions of the outer problems enter at the boundary conditions, Eqs. (5.174) and (5.175).

Cess solved this problem for Pr = 1 and found that

$$F_0'(0) = 1.23 \qquad G_0(0) = 1.01$$

and

$$\theta_0'(0) = -0.567 \qquad \theta_1'(0) = -0.072 \qquad \theta_2'(0) = 0.091$$

One physical quantity of interest is the ratio

$$\begin{aligned}\frac{\mathrm{Nu}}{\mathrm{Gr}^{1/4}} &= -\frac{\xi^{1/2}}{\sqrt{2}(\theta_w - 1)}\left(\frac{\partial\theta}{\partial\bar{\tau}}\right)_{\bar{\tau}=0} \\ &= -\frac{1}{\sqrt{2}}\left[\theta_0'(0) + \theta_1'(0)\Gamma\xi^{1/3} + \theta_2'(0)\Gamma^2\xi^{2/3} + \cdots\right] \\ &= 0.401 + 0.051\Gamma\xi^{1/3} - 0.064\Gamma^2\xi^{2/3} + \cdots \end{aligned} \tag{5.176}$$

5.6 NATURAL CONVECTION AT HIGH PRANDTL NUMBER

We will now consider the natural convection between a semi-infinite flat plate and its surrounding fluid. This problem becomes singular for high Prandtl number since the factor 1/Pr appears in the term with the highest-order derivative. The details of the analysis will be presented here, following the work of Kuiken (1968a).

The boundary layer equations for this flow are

$$\frac{\partial u}{\partial x} + \frac{\partial v}{\partial y} = 0 \tag{5.177}$$

$$\rho\left(u\frac{\partial u}{\partial x} + v\frac{\partial u}{\partial y}\right) = \mu\frac{\partial^2 u}{\partial y^2} + \rho g\beta(T - T_\infty) \tag{5.178}$$

$$u\frac{\partial T}{\partial x} + v\frac{\partial T}{\partial y} = \alpha\frac{\partial^2 T}{\partial y^2} \tag{5.179}$$

and subject to the boundary conditions

$$y = 0 \qquad u = v = 0 \qquad T = T_w \tag{5.180}$$

$$y = \infty \qquad u = 0 \qquad T = T_\infty \tag{5.181}$$

Let us introduce the transformation

$$\eta = \frac{y}{x}\left(\frac{\mathrm{Gr}_x}{4}\right)^{1/4}$$

$$\psi = 2\sqrt{2}\nu(\mathrm{Gr}_x)^{1/4} f(\eta) \tag{5.182}$$

$$\frac{T-T_\infty}{T_w - T_\infty} = \phi(\eta)$$

where

$$u = \frac{\partial \psi}{\partial y} \qquad v = -\frac{\partial \psi}{\partial x} \tag{5.183}$$

and

$$\mathrm{Gr}_x = \frac{g\beta x^3 (T_w - T_\infty)}{\nu^2} \tag{5.184}$$

Equations (5.177)-(5.179) and the boundary conditions in Eqs. (5.180) and (5.181) become

$$\frac{d^3 f}{d\eta^3} + 3f\frac{d^2 f}{d\eta^2} - 2\left(\frac{df}{d\eta}\right)^2 + \phi = 0 \tag{5.185}$$

$$\epsilon^2 \frac{d^2\phi}{d\eta^2} + 3f\frac{d\phi}{d\eta} = 0 \tag{5.186}$$

and

$$\eta = 0 \qquad f(0) = \frac{df(0)}{d\eta} = 0 \qquad \phi(0) = 1 \tag{5.187}$$

$$\eta = \infty \qquad \frac{df(\infty)}{d\eta} = \phi(\infty) = 0 \tag{5.188}$$

where the perturbation parameter is defined as

$$\epsilon = \frac{1}{\sqrt{\mathrm{Pr}}} \tag{5.189}$$

For the inner layer, the following stretching transformation is introduced

$$f = \mathrm{Pr}^{-3/4} F(\xi)$$

$$\phi = \theta(\xi) \tag{5.190}$$

$$\eta = \mathrm{Pr}^{-1/4}\, \xi$$

to recast Eqs. (5.185) and (5.186) into the form

$$\frac{d^3 F}{d\xi^3} + \theta + \epsilon^2\left[3F\frac{d^2 F}{d\xi^2} - 2\left(\frac{dF}{d\xi}\right)^2\right] = 0 \tag{5.191}$$

$$\frac{d^2\theta}{d\xi^2} + 3F\frac{d\theta}{d\xi} = 0 \tag{5.192}$$

For later use, we have

$$\psi_{\text{inner}} = 4\nu c x^{3/4} \epsilon^{3/2} F(\xi) \tag{5.193}$$

$$T_{\text{inner}} = T_\infty + (T_w - T_\infty)\theta(\xi) \tag{5.194}$$

$$u_{\text{inner}} = 4\nu c^2 x^{1/3} \epsilon \frac{dF}{d\xi} \tag{5.195}$$

where
$$c = \left[\frac{\bar{g}\beta(T_w - T_\infty)}{4\nu^2}\right]^{1/4} \tag{5.196}$$

For the outer region, Eq. (5.186) becomes

$$\frac{d\phi}{d\eta} = 0 \tag{5.197}$$

as Pr becomes very large. Equation (5.197) shows that, in the outer layer, $\phi =$ constant. To satisfy the outer boundary condition, $\phi(\infty) = 0$, this constant must be zero. Therefore, in the outer layer

$$\phi = 0 \tag{5.198}$$

For the momentum equation in the outer layer, let us introduce the transformation

$$\begin{aligned} f &= \mathrm{Pr}^{-1/4}\, G(\xi) \\ \eta &= \xi\, \mathrm{Pr}^{1/4} \end{aligned} \tag{5.199}$$

Equation (5.185) becomes

$$\frac{d^3G}{d\xi^3} + 3G\frac{d^2G}{d\xi^2} - 2\left(\frac{dG}{d\xi}\right)^2 = 0 \tag{5.200}$$

For matching purposes, we have

$$\psi_{\text{outer}} = 4\nu c x^{3/4} \epsilon^{1/2} G(\xi) = 4\nu c x^{3/4} \epsilon^{1/2} G(\epsilon\,\xi) \tag{5.201}$$

Next, we assume, for the inner layer, an expansion of the form

$$F(\xi) = F_0(\xi) + \epsilon F_1(\xi) + \epsilon^2 F_2(\xi) + O(\epsilon^3) \tag{5.202}$$

$$\theta(\xi) = \theta_0(\xi) + \epsilon\theta_1(\xi) + \epsilon^2\theta_2(\xi) + O(\epsilon^3) \tag{5.203}$$

Upon substituting Eqs. (5.202) and (5.203) into Eqs. (5.191) and (5.192) and separating the successive perturbations, we get

$$\epsilon^0: \quad \frac{d^3F_0}{d\xi^3} + \theta_0 = 0 \tag{5.204}$$

$$\frac{d^2\theta_0}{d\xi^2} + 3F_0\frac{d\theta_0}{d\xi} = 0 \tag{5.205}$$

$$\epsilon^1: \quad \frac{d^3F_1}{d\xi^3} + \theta_1 = 0 \tag{5.206}$$

$$\frac{d^2\theta_1}{d\xi^2} + 3F_0\frac{d\theta_1}{d\xi} + 3F_1\frac{d\theta_0}{d\xi} = 0 \tag{5.207}$$

$$\epsilon^2: \quad \frac{d^2F_2}{d\xi^3} + \theta_2 + 3F_0\frac{d^2F_0}{d\xi^2} - 2\left(\frac{dF_0}{d\xi}\right)^2 = 0 \tag{5.208}$$

$$\frac{d^2\theta_2}{d\xi^2} + 3F_0\frac{d\theta_2}{d\xi} + 3\frac{d\theta_0}{d\xi}F_2 + 3F_1\frac{d\theta_1}{d\xi} = 0 \tag{5.209}$$

For the outer layer, the following expansion is introduced

$$G(\xi) = G_0(\xi) + \epsilon G_1(\xi) + \epsilon^2 G_2(\xi) + O(\epsilon^3) \tag{5.210}$$

Upon substituting Eq. (5.210) into Eq. (5.200) and separating the successive perturbations, we get

$$\epsilon^0: \quad \frac{d^2G_0}{d\xi^3} + 3G_0\frac{d^2G_0}{d\xi^2} - 2\left(\frac{dG_0}{d\xi}\right)^2 = 0 \tag{5.211}$$

$$\epsilon^1: \quad \frac{d^3G_1}{d\xi^3} + 3G_0\frac{d^2G_1}{d\xi^2} - 4\frac{dG_0}{d\xi}\frac{dG_1}{d\xi} + 3\frac{d^2G_0}{d\xi^2}G_1 = 0 \tag{5.212}$$

$$\epsilon^2: \quad \frac{d^3G_2}{d\xi^3} + 3G_0\frac{d^2G_2}{d\xi^2} - 4\frac{dG_0}{d\xi}\frac{dG_2}{d\xi} + 3\frac{d^2G_0}{d\xi^2}G_2 + 3G_1\frac{d^2G_1}{d\xi^2} - 2\left(\frac{dG_1}{d\xi}\right)^2 = 0 \tag{5.213}$$

Net, let us consider the boundary conditions necessary for the solution of Eqs. (5.204)-(5.213).

For the inner expansion, the inner boundary conditions must be satisfied. We therefore have

$$F_0(0) = F_1(0) = F_2(0) = 0 \tag{5.214a, b, c}$$

$$\frac{dF_0(0)}{d\xi} = \frac{dF_1(0)}{d\xi} = \frac{dF_2(0)}{d\xi} = 0 \tag{5.215a, b, c}$$

$$\theta_0(0) = 1 \qquad \theta_1(0) = \theta_2(0) = 0 \tag{5.216a, b, c}$$

For the outer expansion, the outer boundary conditions must be satisfied. We therefore have

$$\frac{dG_0(\infty)}{d\xi} = \frac{dG_1(\infty)}{d\xi} = \frac{dG_2(\infty)}{d\xi} = 0 \tag{5.217a, b, c}$$

Also, since $\phi = 0$ in the outer layer, it implies that

$$\theta_0(\infty) = \theta_1(\infty) = \theta_2(\infty) = 0 \tag{5.218a, b, c}$$

For the solution of the boundary value problems, nine additional boundary conditions are needed. They will be obtained by matching the two solutions. It is required that

$$\lim_{\xi\to\infty} \psi_{\text{inner}} = \lim_{\xi\to 0} \psi_{\text{outer}} \tag{5.219}$$

which, in terms of inner variables, can be written as

$$\lim_{\xi\to\infty} \epsilon F(\xi) = \lim_{\xi\to\infty} G(\epsilon\xi) \tag{5.220}$$

In terms of the expansions, Eqs. (5.202), (5.203), and (5.210), the condition in Eq. (5.220) becomes

$$\lim_{\xi\to\infty} \epsilon[F_0(\xi) + \epsilon F_1(\xi) + \epsilon^2 F_2(\xi) + O(\epsilon^3)]$$

$$= \lim_{\xi\to 0} [G_0(\epsilon\xi) + \epsilon G_1(\epsilon\xi) + \epsilon^2 G_2(\epsilon\xi) + O(\epsilon^3)] \tag{5.221}$$

Based on a general behavior of the boundary layer, namely, as ξ approaches infinity, the asymptotic behavior of boundary layer solutions generally gives the stream function as a polynomial in ξ plus terms of exponentially small order. For this reason let us expand

$$F_0(\xi) = a_{00} + a_{10}\xi + a_{20}\xi^2 + a_{30}\xi^3 + \cdots \tag{5.222}$$

$$F_1(\xi) = a_{01} + a_{11}\xi + a_{21}\xi^2 + a_{31}\xi^3 + \cdots \tag{5.223}$$

$$F_2(\xi) = a_{02} + a_{12}\xi + a_{22}\xi^2 + a_{32}\xi^3 + \cdots \tag{5.224}$$

where

$$a_{0n} = F_n(\infty)\,,\ a_{1n} = \frac{dF_n(\infty)}{d\xi}\,,\ a_{2n} = \frac{1}{2}\frac{d^2F_n(\infty)}{d\xi^2}\,,\ a_{3n} = \frac{1}{6}\frac{d^3F_n(\infty)}{d\xi^3}\,,\ \cdots \tag{5.225}$$

Also, since ψ_{outer} vanishes at the wall $\xi = 0$, the G_n have expansions of the form

$$G_n(\xi) = b_{0n} + b_{1n}\xi + b_{2n}\xi^2 + b_{3n}\xi^3 + \cdots \tag{5.226}$$

where

$$b_{0n} = G_n(0)\,,\ b_{1n} = \frac{dG_n(0)}{d\xi}\,,\ b_{2n} = \frac{1}{2}\frac{d^2G_n(0)}{d\xi^2}\,,\ b_{3n} = \frac{1}{6}\frac{d^3G_n(0)}{d\xi^3}\,,\ \cdots \tag{5.227}$$

The expansions, Eqs. (5.222)-(5.224), and (5.226), can now be substituted into Eq. (5.221) and rearranged to give

$$\{b_{00}\} + \{b_{01} - a_{00}\}\epsilon + \{b_{10} - a_{10}\}\epsilon\xi + \{-a_{20}\}\epsilon\xi^2 + \{b_{02} - a_{01}\}\epsilon^2$$
$$+ \{b_{11} - a_{11}\}\epsilon^2\xi + \{b_{20} - a_{21}\}\epsilon^2\xi^2 + \{b_{03} - a_{02}\}\epsilon^3 + \{b_{12} - a_{12}\}\epsilon^3\xi$$
$$+ \{b_{21} - a_{22}\}\epsilon^3\xi^2 + \{b_{30} - a_{32}\}\xi^3 + \cdots = 0 \qquad (5.228)$$

By setting the coefficients equal to zero, we get

$$b_{00} = 0: \quad G_0(0) = 0 \qquad (5.229)$$

$$b_{01} - a_{00} = 0: \quad G_1(0) = F_0(\infty) \qquad (5.230)$$

$$b_{10} - a_{10} = 0: \quad \frac{dG_0(0)}{d\xi} = \frac{dF_0(\infty)}{d\xi} \qquad (5.231)$$

$$a_{20} = 0: \quad \frac{d^2F_0(\infty)}{d\xi^2} = 0 \qquad (5.232)$$

$$b_{02} - a_{01} = 0: \quad G_2(0) = F_1(\infty) \qquad (5.223)$$

$$b_{11} - a_{11} = 0: \quad \frac{dG_1(0)}{d\xi} = \frac{dF_1(\infty)}{d\xi} \qquad (5.234)$$

$$b_{20} = a_{21} = 0: \quad \frac{d^2G_0(0)}{d\xi^2} = \frac{d^2F_1(\infty)}{d\xi^2} \qquad (5.235)$$

$$b_{03} - a_{02} = 0: \quad G_3(0) = F_2(\infty) \qquad (5.236)$$

$$b_{12} - a_{12} = 0: \quad \frac{dG_2(0)}{d\xi} = \frac{dF_2(\infty)}{d\xi} \qquad (5.237)$$

$$b_{21} - a_{22} = 0: \quad \frac{d^2G_1(0)}{d\xi^2} = \frac{d^2F_2(\infty)}{d\xi^2} \qquad (5.238)$$

$$b_{30} - a_{32} = 0: \quad \frac{d^3G_0(0)}{d\xi^3} = \frac{d^3F_2(\infty)}{d\xi^3} \qquad (5.239)$$

The solution of the boundary value problems of successive perturbations can be summarized as follows:

1. Equations (5.204) and (5.205) are solved, subject to the boundary conditions, Eqs. (5.214*a*), (5.215*a*), (5.216*a*), (5.218*a*), and (5.232). This will give $F_0(\xi)$ and $\theta_0(\xi)$.
2. Equation (5.211) is solved, subject to the boundary conditions, Eqs. (5.217*a*), (5.229), and (5.231). This will give $G_0(\xi)$.
3. Equations (5.206) and (5.207) are solved, subject to the boundary conditions, Eqs. (5.214*b*), (5.215*b*), (5.216*b*), (5.218*b*), and (5.235). This will give $F_1(\xi)$ and $\theta_1(\xi)$.
4. Equation (5.212) is solved, subject to the boundary conditions, Eqs. (5.217*b*), (5.230), and (5.234). This will give $G_1(\xi)$.

5. Equations (5.208) and (5.209) are solved, subject to the boundary conditions, Eqs. (5.214*c*), (5.215*c*), (5.216*c*), (5.218*c*), and (5.238). This will give $F_2(\xi)$ and $\theta_2(\xi)$.
6. Equation (5.213) is solved, subject to the boundary conditions, Eqs. (5.217*c*), (5.233), and (5.238). This will give $G_2(\xi)$.

It should be noted that the condition in Eq. (5.236) is not used since it relates to higher-order perturbation. The condition in Eq. (5.239) is satisfied as shown below.

From Eq. (5.208) we get

$$\frac{d^3F_2(\infty)}{d\xi^3} + \theta_2(\infty) + 3F_0(\infty)\frac{d^2F_0(\infty)}{d\xi^2} - 2\left[\frac{dF_0(\infty)}{d\xi}\right]^2 = 0 \tag{5.240}$$

Using Eqs. (5.218*c*), (5.232), and (5.231), Eq. (5.240) becomes

$$\frac{d^3F_2(\infty)}{d\xi^3} = 2\left[\frac{dG_0(0)}{d\xi}\right]^2 \tag{5.241}$$

From Eq. (5.211) we have

$$\frac{d^3G_0(0)}{d\xi^3} + 3G_0(0)\frac{d^2G_0(0)}{d\xi^2} - 2\left[\frac{dG_0(0)}{d\xi}\right]^2 = 0 \tag{5.242}$$

Using Eq. (5.229), Eq. (5.242) becomes

$$\frac{d^3G_0(0)}{d\xi^3} = 2\left[\frac{dG_0(0)}{d\xi}\right]^2 \tag{5.243}$$

Equations (5.241) and (5.243) show that the condition in Eq. (5.239) is satisfied.

Solutions of the first three perturbation equations are given by Kuiken (1968a) and some important numerical data are tabulated in Table 5.2.

It can be proved that the Nusselt number is given by

$$\frac{\mathrm{Nu}_x}{\mathrm{Gr}_x^{1/4}} = -\sqrt{2}\,\mathrm{Pr}^{1/4}\left[\frac{d\theta_0(0)}{d\xi} + \mathrm{Pr}^{-1/2}\frac{d\theta_1(0)}{d\xi} + \mathrm{Pr}^{-1}\frac{d\theta_2(0)}{d\xi} + \cdots\right]$$

where the Nusselt number and the Grashof number are given by

$$\mathrm{Nu}_x = \frac{hx}{k} = \frac{x}{T_w - T_\infty}\frac{\partial T(x,0)}{\partial y}$$

$$\mathrm{Gr}_x = \frac{g\beta(T_w - T_\infty)x^3}{\nu^2}$$

Table 5.2 Sample solutions

i	$d^2F_i(0)/d\xi^2$	$d\theta_i(0)/d\xi$	$G_i(\infty)$
0	0.824516	−0.710989	0.429209
1	−0.306698	0.186442	0.021623
2	−0.224248	−0.067251	0.071661

5.7 CONVECTION IN A PIPE

In this section, we discuss a variation of the method of matched asymptotic expansion, known as the method of boundary-layer correction. The terminology of "boundary-layer" is not to be confused with the thin layer of fluid close to the surface in the boundary-layer theory. Rather, it is the inner layer referred to in the last five sections.

To illustrate the details of this method, let us consider the problem of a heat generating fluid in an insulated pipe. The temperature of the fluid is assumed to be uniform at a given cross-section and is a function of x only. The coordinate x is the direction along the axis of the pipe with $x = 0$ at the entrance of the pipe. For simplicity, the velocity is assumed to be uniform over the cross-section (i.e., slug flow). The heat generation is assumed to be proportional to the temperature. Under these assumptions, the energy equation can be written as

$$\frac{1}{\Pr}\frac{d^2\theta}{d\bar{x}^2} + \operatorname{Re}\frac{d\theta}{d\bar{x}} + \beta\theta = 0 \tag{5.244}$$

subject to the boundary conditions

$$\bar{x} = 0 \qquad \theta(0) = 0 \tag{5.245}$$

$$\bar{x} = 1 \qquad \theta(1) = 1 \tag{5.246}$$

where the dimensionless quantities are related to their physical counterparts by

$$\bar{x} = 1 - \frac{x}{L} \qquad \theta = \frac{T - T_L}{T_0 - T_L} \tag{5.247}$$

and β is the dimensionless parameter related to the heat generation term.

Equation (5.244) and its boundary conditions can be written as

$$\epsilon\frac{d^2\theta}{d\bar{x}^2} + \operatorname{Re}\frac{d\theta}{d\bar{x}} + \beta\theta = 0 \tag{5.248}$$

$$\bar{x} = 0 \qquad \theta(0) = 0 \tag{5.249}$$

$$\bar{x} = 1 \qquad \theta(1) = 1 \tag{5.250}$$

where

$$\epsilon = \frac{1}{\mathrm{Pr}} \tag{5.251}$$

is the perturbation parameter. We are looking for the solution of Eq. (5.248) for large Prandtl numbers.

The first step in the method of boundary-layer correction is to obtain the *outer* solution. An outer solution is obtained by dropping the first term of Eq. (5.248). Let $f(\bar{x}, \epsilon)$ be the outer solution, then we get

$$\mathrm{Re}\frac{df}{d\bar{x}} + \beta f = 0 \tag{5.252}$$

$$f(1) = 1 \tag{5.253}$$

We will add to $f(\bar{x}, \epsilon)$ a boundary-layer correction, $\xi(\tau, \epsilon)$, where

$$\tau = \frac{\bar{x}}{\epsilon} \tag{5.254}$$

to the outer solution such that

$$\theta(\bar{x}, \epsilon) = f(\bar{x}, \epsilon) + \xi(\tau, \epsilon) \tag{5.255}$$

Using Eq. (5.255) in Eq. (5.248), we have

$$\epsilon\left(\frac{d^2f}{d\bar{x}^2} + \frac{1}{\epsilon^2}\frac{d^2\xi}{d\tau^2}\right) + \mathrm{Re}\left(\frac{df}{d\bar{x}} + \frac{1}{\epsilon}\frac{d\xi}{d\tau}\right) + \beta(f + \xi) = 0$$

or, multiplying by ϵ,

$$\epsilon^2\frac{d^2f}{d\bar{x}^2} + \frac{d^2\xi}{d\tau^2} + \mathrm{Re}\,\epsilon\frac{df}{d\bar{x}} + \mathrm{Re}\frac{d\xi}{d\tau} + \beta\epsilon(f + \xi) = 0 \tag{5.256}$$

Next, we subtract Eq. (5.252) from Eq. (5.256) which gives

$$\frac{d^2\xi}{d\tau^2} + \mathrm{Re}\frac{d\xi}{d\tau} + \beta\epsilon\xi = -\epsilon^2\frac{d^2f}{d\bar{x}^2} \tag{5.257}$$

The initial condition is

$$\tau = 0 \qquad \xi(0, \epsilon) = -f(0, \epsilon) \tag{5.258}$$

Equation (5.257), subject to the initial condition in Eq. (5.258), gives the boundary correction term.

The solution of Eq. (5.252) is

$$f = \exp\left[\frac{\beta}{\mathrm{Re}}(1 - \bar{x})\right] \tag{5.259}$$

which is the outer solution.

Substituting the outer solution into Eq. (5.257), we get

$$\frac{d^2\xi}{d\tau^2}+\mathrm{Re}\,\frac{d\xi}{d\tau}+\beta\epsilon\xi=-\epsilon^2\,\frac{\beta^2}{\mathrm{Re}^2}\exp\left[\frac{\beta}{\mathrm{Re}}(1-\bar{x})\right] \tag{5.260}$$

To solve Eq. (5.260), let us expand $\xi(\tau,\epsilon)$ as

$$\xi(\tau,\epsilon)=\xi_0(\tau)+\epsilon\xi_1(\tau) \tag{5.261}$$

By substituting the perturbation series, Eq. (5.261), into Eq. (5.260), the different perturbation equations are

$$\epsilon^0:\quad \frac{d^2\xi_0}{d\tau^2}+\mathrm{Re}\,\frac{d\xi_0}{d\tau}=0 \tag{5.262}$$

$$\xi_0(0)=-e^{\beta/\mathrm{Re}} \tag{5.263}$$

$$\epsilon^1:\quad \frac{d^2\xi_1}{d\tau^2}+\mathrm{Re}\,\frac{d\xi_1}{d\tau}=-\beta\xi_0 \tag{5.264}$$

$$\xi_1(0)=0 \tag{5.265}$$

Since $\xi(\tau,\,\epsilon)$ is the "boundary" correction, the boundary condition for large τ should be

$$\tau=\infty\qquad \xi(\infty,\epsilon)=0 \tag{5.266}$$

which gives

$$\begin{aligned}\xi_0(\infty)&=0\\ \xi_1(\infty)&=0\end{aligned} \tag{5.267}$$

The solutions for ξ_0 and ξ_1 are

$$\xi_0(\tau)=-\exp\left(\frac{\beta}{\mathrm{Re}}-\mathrm{Re}\,\tau\right) \tag{5.268}$$

$$\xi_1(\tau)=-\frac{\beta}{\mathrm{Re}}\exp\left(\frac{\beta}{\mathrm{Re}}-\mathrm{Re}\,\tau\right) \tag{5.269}$$

Thus the solution of Eq. (5.244), subject to the boundary conditions, Eqs. (5.245) and (5.246), is

$$\theta(\bar{x})=\exp\left[\frac{\beta}{\mathrm{Re}}(1-\bar{x})\right]-\left[\exp\left(\frac{\beta}{\mathrm{Re}}-\frac{\mathrm{Re}\,\bar{x}}{\epsilon}\right)\left(1+\frac{\beta\bar{x}}{\mathrm{Re}}\right)\right] \tag{5.270}$$

PROBLEMS

5.1 Consider the steady state heat transfer in an axisymmetric stagnation flow on an infinite circular cylinder. The boundary-layer equations are similar and are transformed to the following system of ordinary differential equations (Gorla, 1976)

$$\eta f''' + f'' + \mathrm{Re}\,[1 + ff'' - (f')^2] = 0$$

$$\eta\theta'' + (1 + \mathrm{Re}\,\mathrm{Pr}\, f)\theta' = 0$$

subject to the boundary conditions

$$f(1) = f'(1) = 0 \qquad \theta(1) = 1$$

$$f'(\infty) = 1 \qquad \theta(\infty) = 0$$

Solve this equation for large Prandtl numbers by the method of matched asymptotic expansion and compare with the results tabulated by Gorla.

5.2 In an analysis of interaction between heat and mass transfer in simultaneous natural convection about an isothermal vertical flat plate, J. Schenk et al. (1976) derived the following system of boundary value problems

$$f''' - 2(f')^2 + 3ff'' + \delta\theta + (1 - \delta)w = 0$$

$$\theta'' + 3\,\mathrm{Pr}\, f\theta' = 0$$

$$w'' + 3\,\mathrm{Sc}\, fw' = 0$$

subject to the boundary conditions

$$f(0) = f'(0) = 0 \qquad \theta(0) = 1 \qquad w(0) = 1$$

$$f'(\infty) = 0 \qquad \theta(\infty) = 0 \qquad w(\infty) = 0$$

where Pr, Sc, and δ are, respectively, the Prandtl number, the Schmidt number, and the ratio of the thermal buoyancy to the total body force. Solve this problem for large Prandtl number and or large Schmidt number and for $\delta = 0.5$.

5.3 In an analysis of the dynamics of a buoyant plume rising above a horizontal line heat source in a transverse, horizontal magnetic field, D. D. Gray (1977) found that similarity solutions exist for the boundary-layer equations and the resulting equations, in terms of the similarity variables, are

$$f''' - \tfrac{4}{5}(f')^2 + \tfrac{12}{5}ff'' + \phi - \tfrac{1}{2}Z_L f' = 0$$

$$\phi'' + \tfrac{12}{5}\mathrm{Pr}\,(f\phi)' = 0$$

subject to the boundary conditions

$$f(0) = f''(0) = 0 \qquad \phi(0) = 1$$

$$f'(\infty) = 0 \qquad \phi(\infty) = 0$$

where Pr is the Prandtl number and Z_L is the Lykoudis number which, physically, is a measure of the ratio of the pondermotive force divided by the square root of the buoyancy force times the inertial force. Numerical solutions are given for $\mathrm{Pr} = 0.01$ to 10 and $Z_L = 0$ to 6.7882. However, for very large Prandtl numbers, the problem becomes singular. Solve the problem by the method of matched asymptotic expansion.

5.4 B. Gebhart (1962) was the first to investigate the effect of viscous dissipation on natural convection. The boundary-layer equations are

$$\frac{\partial u}{\partial x} + \frac{\partial v}{\partial y} = 0$$

$$\rho\left(u\,\frac{\partial u}{\partial x} + v\,\frac{\partial u}{\partial y}\right) = \alpha\,\frac{\partial^2 u}{\partial y^2} \pm \rho g \beta \theta$$

$$u\,\frac{\partial \theta}{\partial x} + v\,\frac{\partial \theta}{\partial y} = \alpha\,\frac{\partial^2 \theta}{\partial y^2} + \frac{\nu}{c_p}\left(\frac{\partial u}{\partial y}\right)^2$$

subject to the boundary conditions

$$y = 0 \qquad u = v = 0 \qquad \theta = T_w - T_\infty$$

$$y = \infty \qquad u = 0 \qquad \theta = 0$$

Gebhart developed perturbation solution of the form

$$u = \frac{\partial \psi}{\partial y} \qquad v = -\,\frac{\partial \psi}{\partial x}$$

$$\phi = \frac{T - T_\infty}{T_w - T_\infty} = \phi_0 \pm 4\epsilon\phi_1 \pm \cdots$$

$$\psi = 2\sqrt{2}\nu(\mathrm{Gr}_x)^{1/4}(f_0 \pm 4\epsilon f_1 \pm \cdots)$$

$$\mathrm{Gr}_x = \frac{g\beta x^3(T_w - T_\infty)}{\nu^2}$$

$$\eta = \frac{y}{x}\left(\frac{\mathrm{Gr}_x}{4}\right)^{1/4}$$

$$\epsilon = \frac{g\beta x}{c_p}$$

Solve this problem for large Prandtl number by the method matched asymptotic expansion by following the inner and outer expansion of Roy (1969).

5.5 In an analysis of the laminar free convection from a nonisothermal cone at low Prandtl numbers, R. G. Hering (1965) simplified the boundary-layer equations by a similarity transformation to a system of nonlinear differential equations

$$\Pr f''' + \left(\frac{7+n}{4}\right) ff'' - \left(\frac{1+n}{2}\right)(f')^2 + \theta = 0$$

$$\theta'' + \left(\frac{7+n}{4}\right) f\theta' - nf'\theta = 0$$

$$\eta = 0 \quad f = f' = 0 \quad \theta = 1$$

$$\eta = \infty \quad f' = 0 \quad \theta = 0$$

As the Prandtl number approaches zero, the fluid behavior approaches that of an inviscid fluid where the buoyancy force just balances the initial force. Using the method of matched asymptotic expansion, solve the problem for small Prandtl numbers.

5.6 In an analysis of a belt type radiator, R. J. Krane et al. (1973) obtained the following boundary problem

$$\epsilon \frac{d^2\theta}{dx^2} - \frac{d\theta}{dx} - \alpha(\theta - \theta_0) = 0 \qquad \left(0 \leqslant x \leqslant \frac{a^-}{l}\right)$$

$$\epsilon \frac{d^2\theta}{dx^2} - \frac{d\theta}{dx} - \theta^4 = 0 \qquad \left(\frac{a^+}{l} \leqslant x \leqslant 1\right)$$

subject to the boundary conditions

$$\theta\left(\frac{a^-}{l}\right) = \theta\left(\frac{a^+}{l}\right)$$

$$\theta(0) = \theta(1)$$

$$\frac{d\theta(a^-/l)}{dx} = \frac{d\theta(a^+/l)}{dx}$$

$$\frac{d\theta(0)}{dx} = \frac{d\theta(1)}{dx}$$

Solve the problem for $\epsilon = 0.277 \times 10^{-6}$ and for

$$\alpha = 10, 25, \text{and } 50$$

$$\frac{a}{l} = 0.04 \quad \theta_D = 0.8854$$

5.7 In an analysis of the temperature distributions on a thin heat shield shell subject to longitudinal conduction, radiation losses, and an arbitrary aerodynamic source loading, H. F. Mueller and N. D. Malmuth (1965) obtained the following boundary value problem

$$\epsilon \frac{d^2\theta}{dX^2} - \theta^4 + q(X) = 0$$

$$X = 0 \qquad \frac{d\theta}{dX} = 0$$

$$X = 1 \qquad \theta = 0$$

Using the method of matched asymptotic expansion, solve the problem for

$$q(X) = 0.37 + 0.48 \sin(\pi X) - 0.15 \cos(2\pi X)$$

and for

$$\epsilon = 0.01, 0.005, \text{ and } 0.0001$$

CHAPTER

SIX

EXTENSION, ANALYSIS, AND IMPROVEMENT OF PERTURBATION SERIES

6.1 INTRODUCTION

Most perturbation expansions that we have discussed so far were terminated at the second or third term. In general, such an early truncation of the series is due to three reasons. First, the truncated expansion is usually sufficiently accurate in the range of ϵ that is of practical interest. Second, as one proceeds to the higher-order perturbation equations, their solutions are increasingly more difficult to obtain. Third, the mere proliferation of terms does not necessarily guarantee an increase in accuracy of the solution; in fact it can sometimes impair the accuracy.

Obviously a truncated expansion can only be valid for a limited range of values of the perturbation quantity. Beyond that, it fails to converge and produces erroneous answers. This situation has for a long time been construed as an inherent limitation of the perturbation technique. Consequently, on many occasions, the technique was abandoned in favor of fully numerical computations using a digital computer.

However, recent years have seen a revival of the perturbation method and there is a renewed interest in its applications. This revival has been kindled by the development of extended perturbation series method. The method involves three steps. First, the set of perturbation equations are programmed for solution on a digital computer so that a large number of terms can be generated. Second, the coefficients of the series are utilized to identify the location and nature of singularities limiting the range of applicability of the series. Based on this knowledge, the final step is to recast the series using one or a combination of devices such as Euler transformation, Shanks transformation, Padé approximants, extraction of singularities, series reversion, etc. The improved series generally has better accuracy and wider range of applicability than the original series.

This chapter is devoted entirely to the aforementioned scheme of extension, analysis, and improvement of perturbation series arising in heat transfer.

6.2 EXTENSION OF SERIES

For simple problems, the extension of a perturbation series is not particularly challenging and can be achieved with hand computation alone. There are several examples of such extension in the fluid mechanics literature, but such attempts in heat transfer have been rare.

For complex problems, hand calculation is not feasible and a digital computer must be employed. Depending upon the problem, there are two possible approaches. First, if the problem is such that a pattern can be established for the sequence of solutions, then one can write a program incorporating this pattern and calculate the terms in sequence. However, in order to establish the solution pattern, it is essential to calculate the first few terms by hand. These hand computations also assist in debugging the program. Second, if the solution pattern is not discernable, one must adopt a fully numerical procedure to solve the sequence of perturbation equations. Though both approaches have been used in heat transfer, we will concentrate exclusively on the latter. For a discussion of the former approach the reader is referred to the literature (Anderson and Reynolds, 1970; Pedroso and Domoto, 1973b; Van Dyke, 1975b; Tsan-Hsing Shih, 1981).

We now consider several examples of extended perturbation series.

6.2.1 Heat Transfer in a Radiating Fin

As an example of hand computation, consider a straight rectangular fin radiating to an environment. For uniform base temperature and insulated tip, the governing equations are (Aziz, 1979)

$$\frac{d^2\theta}{dX^2} - \epsilon(\theta^4 - \theta_s^4) = 0 \tag{6.1}$$

$$X = 0 \quad \frac{d\theta}{dX} = 0$$
$$X = 1 \quad \theta = 1 \tag{6.2}$$

where

$$\theta = \frac{T}{T_b} \qquad \theta_s = \frac{T_s}{T_b} \qquad X = \frac{x}{L} \qquad \epsilon = \frac{2E\sigma T_b^3 L^2}{kw} \tag{6.3}$$

and T_b = base temperature
T_s = effective sink temperature
L = fin length
w = fin thickness
k = thermal conductivity
E = emissivity
σ = Stefan-Boltzmann constant
x = axial distance measured from fin tip

For simplicity, assume $\theta_s = 0$. Assuming a regular perturbation expansion, we have

$$\theta = \sum_{n=0}^{\infty} \epsilon^n \theta_n \tag{6.4}$$

Substituting Eq. (6.4) into Eqs. (6.1) and (6.2), the sequence of linear equations appear as

$$\epsilon^0: \quad \frac{d^2\theta_0}{dX^2} = 0 \tag{6.5}$$

$$\begin{aligned} X = 0 \qquad & \frac{d\theta_0}{dX} = 0 \\ X = 1 \qquad & \theta_0 = 1 \end{aligned} \tag{6.6}$$

$$\epsilon^1: \quad \frac{d^2\theta_1}{dX^2} = a_0 \tag{6.7}$$

$$\begin{aligned} X = 0 \qquad & \frac{d\theta_1}{dX} = 0 \\ X = 1 \qquad & \theta_1 = 0 \end{aligned} \tag{6.8}$$

$$\epsilon^i: \quad \frac{d^2\theta_i}{dX^2} = a_{i-1} \qquad i > 1 \tag{6.9}$$

$$\begin{aligned} X = 0 \qquad & \frac{d\theta_i}{dX} = 0 \\ X = 1 \qquad & \theta_i = 0 \end{aligned} \tag{6.10}$$

where $$a_0 = \theta_0^4 \qquad a_m = \frac{1}{m\theta_0} \sum_{j=1}^{m} (5j - m) a_{m-j} \theta_j \tag{6.11}$$

Table 6.1 Coefficients a_m

$$a_0 = \theta_0^4$$

$$a_1 = 4\theta_1\theta_0^3$$

$$a_2 = 4\theta_2\theta_0^3 + 6\theta_1^2\theta_0^2$$

$$a_3 = 4\theta_3\theta_0^3 + 12\theta_2\theta_1\theta_0^2 + 4\theta_1^3\theta_0$$

$$a_4 = \theta_1^4 + 4\theta_4\theta_0^3 + 12\theta_3\theta_1\theta_0^2 + 6\theta_2^2\theta_0^2 + 12\theta_2\theta_1^2\theta_0$$

$$a_5 = 4\theta_5\theta_0^3 + 12\theta_4\theta_1\theta_0^2 + 12\theta_3\theta_2\theta_0^2 + 12\theta_3\theta_1^2\theta_0 + 12\theta_2^2\theta_1\theta_0 + 4\theta_2\theta_1^3$$

$$\begin{aligned} a_6 = {} & 4\theta_6\theta_0^3 + 12\theta_5\theta_1\theta_0^2 + 12\theta_4\theta_2\theta_0^2 + 12\theta_4\theta_1^2\theta_0 + 6\theta_3^2\theta_0^2 + 24\theta_3\theta_2\theta_1\theta_0 \\ & + 4\theta_3\theta_1^3 + 4\theta_2^3\theta_0 + 6\theta_2^2\theta_1^2 \end{aligned}$$

$$\begin{aligned} a_7 = {} & 4\theta_7\theta_0^3 + 12\theta_6\theta_1\theta_0^2 + 12\theta_5\theta_2\theta_0^2 + 12\theta_5\theta_1^2\theta_0 + 12\theta_4\theta_3\theta_0^2 + 4\theta_4\theta_1^3 \\ & + 24\theta_0\theta_1\theta_2\theta_4 + 12\theta_3\theta_2^2\theta_0 + 12\theta_3\theta_2\theta_1^2 + 12\theta_3^2\theta_1\theta_0 + 4\theta_2^3\theta_1 \end{aligned}$$

$$\begin{aligned} a_8 = {} & \theta_2^4 + 4\theta_8\theta_0^3 + 12\theta_7\theta_1\theta_0^2 + 12\theta_6\theta_2\theta_0^2 + 12\theta_6\theta_1^2\theta_0 + 12\theta_5\theta_3\theta_0^2 + 24\theta_5\theta_2\theta_1\theta_0 \\ & + 4\theta_5\theta_1^3 + 6\theta_4^2\theta_0^2 + 12\theta_4\theta_2^2\theta_0 + 12\theta_4\theta_2\theta_1^2 + 24\theta_4\theta_3\theta_1\theta_0 + 12\theta_3^2\theta_2\theta_0 \\ & + 12\theta_3\theta_2^2\theta_1 + 6\theta_3^2\theta_1^2 \end{aligned}$$

$$\begin{aligned} a_9 = {} & 4\theta_9\theta_0^3 + 12\theta_8\theta_1\theta_0^2 + 12\theta_7\theta_2\theta_0^2 + 12\theta_7\theta_1^2\theta_0 + 12\theta_6\theta_3\theta_0^2 + 24\theta_6\theta_2\theta_1\theta_0 \\ & + 4\theta_6\theta_1^3 + 12\theta_5\theta_4\theta_0^2 + 24\theta_5\theta_3\theta_1\theta_0 + 12\theta_5\theta_2\theta_1^2 + 12\theta_5\theta_2^2\theta_0 + 12\theta_4^2\theta_1\theta_0 \\ & + 24\theta_4\theta_3\theta_2\theta_0 + 12\theta_4\theta_3\theta_1^2 + 12\theta_4\theta_2^2\theta_1 + 4\theta_3^3\theta_0 + 12\theta_3^2\theta_2\theta_1 + 4\theta_3\theta_2^3 \end{aligned}$$

Because of the occurrence of the radiation term θ^4, we give in Table 6.1 the coefficients a_0 through a_9 to facilitate computations. The solutions of Eqs. (6.5)-(6.10) involve elementary integration, though the length of the higher-order terms increases rapidly. To conserve space, we quote the first five terms but give the eleven-term series for the tip temperature $\theta(0)$. These are

$$\theta_0 = 1 \tag{6.12}$$

$$\theta_1 = \frac{1}{2}(X^2 - 1) \tag{6.13}$$

$$\theta_2 = \frac{1}{6}(X^4 - 6X^2 + 5) \tag{6.14}$$

$$\theta_3 = \frac{1}{180}(13X^6 - 105X^4 + 435X^2 - 343) \tag{6.15}$$

$$\theta_4 = \frac{1}{15120}(438X^8 - 5460X^6 + 27930X^4 - 99204X^5 + 76251) \tag{6.16}$$

$$\begin{aligned} \theta(0) = {} & 1.000000 - 0.500000\epsilon + 0.833333\epsilon^2 - 1.905555\epsilon^3 + 5.043055\epsilon^4 \\ & + 14.514290\epsilon^5 + 44.158916\epsilon^6 - 140.236100\epsilon^7 + 457.870867\epsilon^8 \\ & - 1526.744970\epsilon^9 + 5175.665448\epsilon^{10} + O(\epsilon^{11}) \end{aligned} \tag{6.17}$$

6.2.2 Transient Conduction into a Variable Conductivity Semi-Infinite Medium

Consider the problem of transient conduction in a semi-infinite medium. For a sudden change in surface temperature and linear increase of thermal conductivity with temperature, the energy equation can be written as

$$\rho C \frac{\partial T}{\partial t} = \frac{\partial}{\partial x} \left\{ k_0 [1 + \alpha(T - T_\infty)] \frac{\partial T}{\partial x} \right\}$$

subject to the boundary conditions

$$T(0, t) = T_0 \qquad T(\infty, t) = T_\infty$$

and the initial condition

$$T(x, 0) = T_\infty$$

where ρ = density
C = specific heat
k_0, α = constants in the expression of the heat conductivity

Other quantities are obvious.

By introducing the similarity transformation

$$\eta = \frac{1}{2} \sqrt{\frac{\rho C}{k_0}} \frac{x}{\sqrt{t}}$$

and

$$\theta = \frac{T - T_\infty}{T_0 - T_\infty}$$

the energy equation can be reduced to

$$\theta'' + 2\eta\theta' + \epsilon(\theta\theta'' + \theta'^2) = 0 \tag{6.18}$$

$$\begin{aligned} \eta &= 0 \qquad \theta = 1 \\ \eta &= \infty \qquad \theta = 0 \end{aligned} \tag{6.19}$$

Assume a regular perturbation expansion of the form in Eq. (6.4), and substitute into Eqs. (6.18) and (6.19) to obtain

$$\sum_{n=0}^{\infty} \epsilon^n \theta_n'' + \epsilon \left[\sum_{n=0}^{\infty} \epsilon^n \theta_n \sum_{n=0}^{\infty} \epsilon^n \theta_n'' + \left(\sum_{n=0}^{\infty} \epsilon^n \theta_n' \right)^2 \right]$$

$$+ 2\eta \sum_{n=0}^{\infty} \epsilon^n \theta_n' = 0 \tag{6.20}$$

Using the theorem for multiplication of power series and rearranging, we have

$$\sum_{n=0}^{\infty} \epsilon^n \theta_n'' + 2\eta \sum_{n=0}^{\infty} \epsilon^n \theta_n' + \sum_{n=0}^{\infty} \epsilon^{n+1}(\theta_0\theta_n'' + \theta_1\theta_{n-1}'' + \cdots + \theta_n\theta_0'')$$

$$+ \sum_{n=0}^{\infty} \epsilon^{n+1}(\theta_0'\theta_n' + \theta_1'\theta_{n-1}' + \cdots + \theta_n'\theta_0') = 0 \tag{6.21}$$

$$\eta = 0 \quad \theta_0 = 1 \quad \theta_n = 0 \quad n = 1, 2, 3, \ldots \tag{6.22}$$

$$\eta = \infty \quad \theta_n = 0 \quad n = 0, 1, 2, 3, \ldots \tag{6.23}$$

We truncate the series at the seventh term. The equations for θ_0 through θ_6 are

$$\epsilon^0: \quad \theta_0'' + 2\eta\theta_0' = 0 \tag{6.24}$$

$$\epsilon^1: \quad \theta_1'' + 2\eta\theta_1' = -(\theta_0\theta_0'' + \theta_0'^2) \tag{6.25}$$

$$\epsilon^2: \quad \theta_2'' + 2\eta\theta_2' = -(\theta_0\theta_1'' + \theta_1\theta_0'' + 2\theta_0'\theta_1') \tag{6.26}$$

$$\epsilon^3: \quad \theta_3'' + 2\eta\theta_3' = -(\theta_0\theta_2'' + \theta_1\theta_1'' + \theta_2\theta_0'' + 2\theta_0'\theta_2' + \theta_1'^2) \tag{6.27}$$

$$\epsilon^4: \quad \theta_4'' + 2\eta\theta_4' = -(\theta_0\theta_3'' + \theta_1\theta_2'' + \theta_2\theta_1'' + \theta_3\theta_0'' + 2\theta_0'\theta_3' + 2\theta_1'\theta_2') \tag{6.28}$$

$$\epsilon^5: \quad \theta_5'' + 2\eta\theta_5' = -(\theta_0\theta_4'' + \theta_1\theta_3'' + \theta_2\theta_2'' + \theta_3\theta_1'' + \theta_4\theta_0'' + 2\theta_0'\theta_4' + 2\theta_1'\theta_3' + \theta_2'^2) \tag{6.29}$$

$$\epsilon^6: \quad \theta_6'' + 2\eta\theta_6' = -(\theta_0\theta_5'' + \theta_1\theta_4'' + \theta_2\theta_3'' + \theta_3\theta_2'' + \theta_4\theta_1'' + \theta_5\theta_0'' + 2\theta_0'\theta_5' + 2\theta_1'\theta_4' + 2\theta_2'\theta_3') \tag{6.30}$$

Equations (6.24)–(6.30) constitute a set of linear, two-point boundary value problems which can be solved by the method of superposition (see Chapter 2 in Na, 1979). A quantity of interest is the dimensionless wall heat flux $-\theta'(0)$. From the numerical results the series for $-\theta'(0)$ follows as

$$-\theta'(0) = \frac{2}{\sqrt{\pi}}(1 - 0.681690\epsilon + 0.627050\epsilon^2 - 0.610747\epsilon^3 + 0.611968\epsilon^4 - 0.624207\epsilon^5 + 0.642933\epsilon^6) \tag{6.31}$$

6.2.3 Boundary-Layer Flow Longitudinal to a Cylinder

This example deals with the boundary-layer flow over a slender circular cylinder when the direction of the main flow is parallel to the axis of the cylinder. The boundary-layer equations can be written as

$$\frac{\partial(ru)}{\partial x}+\frac{\partial(rv)}{\partial r}=0$$

$$u\frac{\partial u}{\partial x}+v\frac{\partial u}{\partial r}=\frac{\nu}{r}\frac{\partial}{\partial r}\left(r\frac{\partial u}{\partial r}\right)$$

subject to the boundary conditions

$$r=R \qquad u=0 \qquad v=0$$

$$r\to\infty \qquad u=U_\infty$$

By introducing the transformation

$$\epsilon=\frac{4}{R}\sqrt{\frac{\nu x}{U_\infty}}$$

$$\eta=\sqrt{\frac{U_\infty}{\nu x}}\,\frac{r^2-R^2}{4R}$$

$$\psi=R\sqrt{\nu U_\infty x}\,f(\epsilon,\eta)$$

the equation of continuity is satisfied identically and the momentum equation becomes

$$(1+\epsilon\eta)f'''+ff''+\epsilon f''=\epsilon\left(f'\frac{\partial f'}{\partial\epsilon}-f''\frac{\partial f}{\partial\epsilon}\right) \tag{6.32}$$

$$f(\epsilon,0)=f'(\epsilon,0)=0 \qquad f'(\epsilon,\infty)=2 \tag{6.33}$$

where the dimensionless quantities f, η, and ϵ are the stream function, similarity variable, and transverse curvature parameter, respectively. A perturbation expansion for f in terms of ϵ was originally devised by Seban and Bond (1951) who calculated the first three terms. Much later, the fourth term was added by Wanous and Sparrow (1965). Here we shall extend it to nine terms. Let

$$f=\sum_{n=0}^{\infty}\epsilon^n f_n \tag{6.34}$$

Then the sequence of perturbation equations following from Eqs. (6.32) and (6.33) is

$$\epsilon^0:\quad f_0'''+f_0f_0''=0 \tag{6.35}$$

$$f_0(0)=f_0'(0)=0 \qquad f_0'(\infty)=2 \tag{6.36}$$

$$\epsilon^n:\quad f_n'''+f_0f_n''+(1+n)f_0''f_n-nf_0'f_n'=r_n \tag{6.37}$$

$$f_n(0)=f_n'(0)=f_n'(\infty)=0 \tag{6.38}$$

where $n = 1, 2, 3, \ldots, 8$ and

$$\begin{aligned}
r_1 &= -\eta f_0''' - f_0'' \\
r_2 &= -\eta f_1''' - f_1'' + (f_1')^2 - 2f_1 f_1'' \\
r_3 &= -\eta f_2''' - f_2'' + 3f_1' f_2' - 2f_1 f_2'' - 3f_2 f_1'' \\
r_4 &= -\eta f_3''' - f_3'' + 4f_1' f_3' + 2(f_2')^2 - 2f_1 f_3'' - 3f_2 f_2'' - 4f_3 f_1'' \\
r_5 &= -\eta f_4''' - f_4'' + 5f_1' f_4' + 5f_2' f_3' - 2f_1 f_4'' - 3f_2 f_3'' - 4f_3 f_2'' - 5f_4 f_1'' \\
r_6 &= -\eta f_5''' - f_5'' + 6f_1' f_5' + 6f_2' f_4' + 3(f_3')^2 - 2f_1 f_5'' - 3f_2 f_4'' - 4f_3 f_3'' \\
&\quad - 5f_4 f_2'' - 6f_5 f_1'' \\
r_7 &= -\eta f_6''' - f_6'' + 7f_1' f_6' + 7f_2' f_5' + 7f_3' f_4' - 2f_1 f_6'' - 3f_2 f_5'' - 4f_3 f_4'' \\
&\quad - 5f_4 f_3'' - 6f_5 f_2'' - 7f_6 f_1'' \\
r_8 &= -\eta f_7''' - f_7'' + 8f_1' f_7' + 8f_2' f_6' + 8f_3' f_5' + 4(f_4')^2 - 2f_1 f_7'' - 3f_2 f_6'' \\
&\quad - 4f_3 f_5'' - 5f_4 f_4'' - 6f_5 f_3'' - 7f_6 f_2'' - 8f_7 f_1''
\end{aligned}$$

The zero-order problem, Eqs. (6.35) and (6.36), is the classical Blasius problem for a flat plate whose solution is well documented. Equations (6.37) and (6.38) constitute a set of linear boundary value problems which can be solved by the method of superposition (Na, 1979).

The shearing stress on the wall is defined by

$$\tau_w = \mu \left(\frac{\partial u}{\partial y}\right)_{y=0}$$

In terms of the similarity variables, the above expression becomes

$$\tau_w = \mu \frac{U_\infty}{4} \sqrt{\frac{U_\infty}{\nu x}}\, f''(\epsilon, 0)$$

or,

$$\frac{\tau_w}{(1/2)\rho U_\infty^2} = \frac{1}{2} \sqrt{\frac{\nu}{U_\infty x}}\, f''(\epsilon, 0)$$

or,

$$C_f (\mathrm{Re}_x)^{1/2} = \tfrac{1}{2} f''(\epsilon, 0) \tag{6.39}$$

We focus attention on the series for $f''(\epsilon, 0)$. From our own numerical results, it follows that

$$\begin{aligned}
f''(\epsilon, 0) = 1.3282 &+ 0.6943\epsilon - 0.1641\epsilon^2 + 0.1016\epsilon^3 - 0.07595\epsilon^4 \\
&+ 0.06129\epsilon^5 - 0.05115\epsilon^6 + 0.04321\epsilon^7 - 0.03647\epsilon^8
\end{aligned} \tag{6.40}$$

where the first four terms are in full agreement with those of Wanous and Sparrow (1965).

6.2.4 Heat Transfer in Falkner-Skan Flow

Unlike the last example, this example involves the solution of both momentum and energy equations. Consider the boundary-layer flow over a wedge when the free stream velocity U_∞ is of the form $U_\infty \sim x^m$ and the surface temperature is constant (but different from the temperature of the main stream). The boundary-layer equations can be written as

$$\frac{\partial u}{\partial x} + \frac{\partial v}{\partial y} = 0$$

$$\rho\left(u\frac{\partial u}{\partial x} + v\frac{\partial u}{\partial y}\right) = \rho U_\infty \frac{dU_\infty}{dx} + \mu\frac{\partial^2 u}{\partial y^2}$$

$$\rho C\left(u\frac{\partial T}{\partial x} + v\frac{\partial T}{\partial y}\right) = k\frac{\partial^2 T}{\partial y^2}$$

subject to the boundary conditions

$$y = 0 \qquad u = 0 \qquad v = 0 \qquad T = T_w$$

$$y = \infty \qquad u = U_\infty \qquad T = T_\infty$$

By introducing the similarity transformation

$$\eta = \sqrt{\frac{m+1}{2}}\,\frac{\bar{y}}{\bar{x}^{(1-m)/2}} \qquad \bar{\psi} = \frac{\psi}{\sqrt{\nu L U_\infty}} = \sqrt{\frac{2}{m+1}}\,\bar{x}^{(1+m)/2} f(\eta)$$

$$\theta = \frac{T - T_\infty}{T_w - T_\infty} = g(\eta)$$

where $\bar{x} = x/L$ and $\bar{y} = \sqrt{\mathrm{Re}}\,y/L$, the boundary-layer equations are transformed to

$$f''' + ff'' + \epsilon[1 - (f')^2] = 0 \tag{6.41}$$

$$g'' + \mathrm{Pr}\, fg' = 0 \tag{6.42}$$

$$f(0) = 0 \qquad f'(0) = 0 \qquad g(0) = 1 \tag{6.43}$$

$$f'(\infty) = 1 \qquad g(\infty) = 0 \tag{6.44}$$

where the dimensionless quantities f, g, and Pr are, respectively, the stream function, temperature, and Prandtl number, and $\epsilon = 2m/(m+1)$. Primes denote differentiation with respect to similarity variable η. Let us expand both f and g as series in ϵ

$$f = \sum_{n=0}^{\infty} \epsilon^n f_n \tag{6.45}$$

$$g = \sum_{n=0}^{\infty} \epsilon^n g_n \tag{6.46}$$

The resulting sequence of perturbation equations up to the eleventh term is

$$\epsilon^0: \quad f_0''' + f_0 f_0'' = 0 \tag{6.47}$$

$$g_0'' + \Pr f_0 g_0' = 0 \tag{6.48}$$

$$f_0(0) = f_0'(0) = 0 \qquad g_0(0) = 1 \tag{6.49}$$

$$f_0'(\infty) = 1 \qquad g_0(\infty) = 0 \tag{6.50}$$

$$\epsilon^n: \quad f_n''' + f_0 f_n'' + f_0'' f_n = r_n \tag{6.51}$$

$$g_n'' + \Pr f_0 g_n' = s_n \tag{6.52}$$

$$f_n(0) = f_n'(0) = 0 \qquad g_n(0) = 0 \tag{6.53}$$

$$f_n'(\infty) = 0 \qquad g_n(\infty) = 0 \tag{6.54}$$

where $n = 1, 2, \ldots, 10$ and

$$\begin{aligned}
r_1 &= (f_0')^2 - 1 \\
r_2 &= 2f_0'f_1' - f_1 f_1'' \\
r_3 &= 2f_0'f_2' + (f_1')^2 - f_1 f_2'' - f_2 f_1'' \\
r_4 &= 2(f_0'f_3' + f_1'f_2') - f_1 f_3'' - f_2 f_2'' - f_3 f_1'' \\
r_5 &= 2(f_0'f_4' + f_1'f_3') + (f_2')^2 - f_1 f_4'' - f_2 f_3'' - f_3 f_2'' - f_4 f_1'' \\
r_6 &= 2(f_0'f_5' + f_1'f_4' + f_2'f_3') - f_1 f_5'' - f_2 f_4'' - f_3 f_3'' - f_4 f_2'' - f_5 f_1'' \\
r_7 &= 2(f_0'f_6' + f_1'f_5' + f_2'f_4') + (f_3')^2 - f_1 f_6'' - f_2 f_5'' - f_3 f_4'' - f_4 f_3'' - f_5 f_2'' \\
&\quad - f_6 f_1'' \\
r_8 &= 2(f_0'f_7' + f_1'f_6' + f_2'f_5' + f_3'f_4') - f_1 f_7'' - f_2 f_6'' - f_3 f_5'' - f_4 f_4'' - f_5 f_3'' \\
&\quad - f_6 f_2'' - f_7 f_1'' \\
r_9 &= 2(f_0'f_8' + f_1'f_7' + f_2'f_6' + f_3'f_5') + (f_4')^2 - f_1 f_8'' - f_2 f_7'' - f_3 f_6'' - f_4 f_5'' \\
&\quad - f_5 f_4'' - f_6 f_3'' - f_7 f_2'' - f_8 f_1'' \\
r_{10} &= 2(f_0'f_9' + f_1'f_8' + f_2'f_7' + f_3'f_6' + f_4'f_5') - f_1 f_9'' - f_2 f_8'' - f_3 f_7'' - f_4 f_6'' \\
&\quad - f_5 f_5'' - f_6 f_4'' - f_7 f_3'' - f_8 f_2'' - f_9 f_1''
\end{aligned}$$

and

$$s_1 = -\Pr f_1 g_0'$$
$$s_2 = -\Pr (f_1 g_1' + f_2 g_0')$$
$$s_3 = -\Pr (f_1 g_2' + f_2 g_1' + f_3 g_0')$$
$$s_4 = -\Pr (f_1 g_3' + f_2 g_2' + f_3 g_1' + f_4 g_0')$$
$$s_5 = -\Pr (f_1 g_4' + f_2 g_3' + f_3 g_2' + f_4 g_1' + f_5 g_0')$$
$$s_6 = -\Pr (f_1 g_5' + f_2 g_4' + f_3 g_3' + f_4 g_2' + f_5 g_1' + f_6 g_0')$$
$$s_7 = -\Pr (f_1 g_6' + f_2 g_5' + f_3 g_4' + f_4 g_3' + f_5 g_2' + f_6 g_1' + f_7 g_0')$$
$$s_8 = -\Pr (f_1 g_7' + f_2 g_6' + f_3 g_5' + f_4 g_4' + f_5 g_3' + f_6 g_2' + f_7 g_1' + f_8 g_0')$$
$$s_9 = -\Pr (f_1 g_8' + f_2 g_7' + f_3 g_6' + f_4 g_5' + f_5 g_4' + f_6 g_3' + f_7 g_2' + f_8 g_1' + f_9 g_0')$$
$$s_{10} = -\Pr (f_1 g_9' + f_2 g_8' + f_3 g_7' + f_4 g_6' + f_5 g_5' + f_6 g_4' + f_7 g_3' + f_8 g_2' + f_9 g_1' + f_{10} g_0')$$

Once again, the zero-order problem defined by Eqs. (6.47)-(6.50) corresponds to the well explored case of the flat plate. Being linear, the rest of the Eqs. (6.51)-(6.54) can be solved by the method of superposition. Since the wall shear stress and the heat transfer rate are related to $f''(0)$ and $g'(0)$ respectively, we quote the series for them. The final results are

$$f''(0) = 0.4696 + 1.29893\epsilon - 1.52215\epsilon^2 + 3.56297\epsilon^3 + 10.6720\epsilon^4 + 36.4617\epsilon^5 - 134.945\epsilon^6 + 526.529\epsilon^7 - 2132.41\epsilon^8 + 8878.32\epsilon^9 - 37762.7\epsilon^{10} \tag{6.55}$$

and for $\Pr = 0.72$,

$$-g'(0) = 0.4181 + 0.2119\epsilon - 0.4344\epsilon^2 + 1.2100\epsilon^3 - 3.9425\epsilon^4 + 14.0958\epsilon^5 - 53.7669\epsilon^6 + 213.8442\epsilon^7 - 877.9190\epsilon^8 + 3692.5437\epsilon^9 - 15829.9031\epsilon^{10} \tag{6.56}$$

6.2.5 Flat Plate Heat Transfer with Variable Freestream Velocity

The variation of freestream velocity with streamwise distance introduces nonsimilar terms in both the momentum and the energy equations. Let us consider the boundary-layer flow of an incompressible fluid when the plate is maintained at a uniform temperature and the free stream velocity U_∞ assumes the form $U_\infty = U_{\infty 0}(1 + \alpha x/L)$. The boundary-layer equations can be written as

$$\frac{\partial u}{\partial x} + \frac{\partial v}{\partial y} = 0$$

$$\rho\left(u\,\frac{\partial u}{\partial x} + v\,\frac{\partial u}{\partial y}\right) = \rho U_\infty\,\frac{dU_\infty}{dx} + \mu\,\frac{\partial^2 u}{\partial y^2}$$

$$\rho C\left(u\,\frac{\partial T}{\partial x} + v\,\frac{\partial T}{\partial y}\right) = k\,\frac{\partial^2 T}{\partial y^2}$$

subject to the boundary conditions

$$y = 0 \qquad u = 0 \qquad v = 0 \qquad T = T_w$$

$$y = \infty \qquad u = U_\infty = U_{\infty 0}\left(1 + \alpha\,\frac{x}{L}\right) \qquad T = T_\infty$$

By introducing the similarity transformation

$$\eta = \bar{y}\,\sqrt{\frac{\bar{U}_\infty}{\bar{x}}} \qquad \epsilon = \bar{x}$$

$$f(\epsilon, \eta) = \frac{\bar{\psi}}{\sqrt{\bar{x}\bar{U}_\infty}} \qquad \theta = \frac{T - T_\infty}{T_w - T_\infty}$$

where

$$\bar{x} = \frac{x}{L} \qquad \bar{y} = \sqrt{\mathrm{Re}_L}\,\frac{y}{L} \qquad \bar{\psi} = \frac{\psi}{\sqrt{\nu L U_\infty}} \qquad \bar{U}_\infty = \frac{U_\infty}{U_{\infty 0}} = 1 + \alpha\epsilon$$

the boundary-layer equations become

$$(1 + \epsilon\alpha)f''' + \left(\frac{1}{2} + \epsilon\alpha\right)ff'' + \epsilon\alpha[1 - (f')^2] = \epsilon(1 + \epsilon\alpha)\left(f'\,\frac{\partial f'}{\partial\epsilon} - f''\,\frac{\partial f}{\partial\epsilon}\right) \tag{6.57}$$

$$\frac{1}{\mathrm{Pr}}(1 + \epsilon\alpha)g'' + \left(\frac{1}{2} + \epsilon\alpha\right)fg' = \epsilon(1 + \epsilon\alpha)\left(f'\,\frac{\partial g}{\partial\epsilon} - g'\,\frac{\partial f}{\partial\epsilon}\right) \tag{6.58}$$

$$f(\epsilon, 0) = f'(\epsilon, 0) \qquad g(\epsilon, 0) = 1 \tag{6.59}$$

$$f'(\epsilon, \infty) = 1 \qquad g(\epsilon, \infty) = 0 \tag{6.60}$$

where f = stream function
g = temperature
α = freestream velocity parameter
Pr = Prandtl number
ϵ = streamwise distance

Primes denote differentiation with respect to similarity variable η. Expanding both f and g according to Eqs. (6.45) and (6.46) and truncating the series at the seventh term, we get the following sequence of equations

$$\epsilon^0: \quad f_0''' + \frac{1}{2} f_0 f_0'' = 0 \tag{6.61}$$

$$\frac{1}{\Pr} g_0'' + \frac{1}{2} f_0 g_0' = 0 \tag{6.62}$$

$$f_0(0) = f_0'(0) = 0 \qquad g_0(0) = 1 \qquad f_0'(\infty) = 1 \qquad g_0(\infty) = 1 \tag{6.63}$$

$$\epsilon^n: \quad f_n''' + \frac{1}{2} f_0 f_n'' + \frac{2n+1}{2} f_0'' f_n - n f_0' f_n' = r_n \tag{6.64}$$

$$\frac{1}{\Pr} g_n'' + \frac{1}{2} f_0 g_n' + \frac{2n+1}{2} g_0' f_n - n f_0' g_n = s_n \tag{6.65}$$

$f_n(0) = f_n'(0) = g_n(0) = f_n(\infty) = g_n(\infty) = 0$, where n O $1, 2, 3, \ldots, 6$ and

$$r_1 = -\alpha f_0''' - \alpha f_0 f_0'' - \alpha + \alpha (f_0')^2$$

$$\begin{aligned} r_2 = {} & -\alpha f_1''' - \tfrac{1}{2} f_1 f_1'' - \alpha(f_0 f_1'' + f_1 f_0'') + 2\alpha f_0' f_1' + (f_1')^2 - f_1 f_1'' \\ & + \alpha(f_1' f_0' - f_1 f_0'') \end{aligned}$$

$$\begin{aligned} r_3 = {} & -\alpha f_2''' - \tfrac{1}{2}(f_1 f_2'' + f_2 f_1'') - \alpha(f_0 f_2'' + f_1 f_1'' + f_2 f_0'') + \alpha[2 f_0' f_2' + (f_1')^2] \\ & + (f_1' f_2' + 2 f_2' f_1') - (f_1 f_2'' + 2 f_2 f_1'') - \alpha f_1 f_0'' \end{aligned}$$

$$\begin{aligned} r_4 = {} & -\alpha f_3''' - \tfrac{1}{2}(f_1 f_3'' + f_2 f_2'' + f_3 f_1'') - \alpha(f_0 f_3'' + f_1 f_2'' + f_2 f_1'' + f_3 f_0'') \\ & + 2\alpha(f_0' f_3' + f_1' f_2') + (f_1' f_3' + 2 f_2' f_2' + 3 f_3' f_1') - (f_1 f_3'' + 2 f_2 f_2'' + 3 f_3 f_1'') \\ & + \alpha(f_1' f_2' + 2 f_2' f_1' + 3 f_3' f_0') - \alpha(f_1 f_2'' + 2 f_2 f_1'' + 3 f_3 f_0'') \end{aligned}$$

$$\begin{aligned} r_5 = {} & -\alpha f_4''' - \tfrac{1}{2}(f_1 f_4'' + f_2 f_3'' + f_3 f_2'' + f_4 f_1'') \\ & - \alpha(f_0 f_4'' + f_1 f_3'' + f_2 f_2'' + f_3 f_1'' + f_4 f_0'') + \alpha[2(f_0' f_4' + f_1' f_3') + (f_2')^2] \\ & + (f_1' f_4' + 2 f_2' f_3' + 3 f_3' f_2' + 4 f_4' f_1') - (f_1 f_4'' + 2 f_2 f_3'' + 3 f_3 f_2'' + 4 f_4 f_1'') \\ & + \alpha(f_1' f_3' + 2 f_2' f_2' + 3 f_3' f_1' + 4 f_4' f_0') \\ & - \alpha(f_1 f_3'' + 2 f_2 f_2'' + 3 f_3 f_1'' + 4 f_4 f_0'') \end{aligned}$$

$$\begin{aligned} r_6 = {} & -\alpha f_5''' - \tfrac{1}{2}(f_1 f_5'' + f_2 f_4'' + f_3 f_3'' + f_4 f_2'' + f_5 f_1'') \\ & - \alpha(f_0 f_5'' + f_1 f_4'' + f_2 f_3'' + f_3 f_2'' + f_4 f_1'' + f_5 f_0'') + 2\alpha(f_0' f_5' + f_1' f_4' + f_2' f_3') \\ & + (f_1' f_5' + 2 f_2' f_4' + 3 f_3' f_3' + 4 f_4' f_2' + 5 f_5' f_1') \\ & - (f_1 f_5'' + 2 f_2 f_4'' + 3 f_3 f_3'' + 4 f_4 f_2'' + 5 f_5 f_1'') \\ & + \alpha(f_1' f_4' + 2 f_2' f_3' + 3 f_3' f_2' + 4 f_4' f_1' + 5 f_5' f_0') \\ & - \alpha(f_1 f_4'' + 2 f_2 f_3'' + 3 f_3 f_2'' + 4 f_4 f_1'' + 5 f_5 f_0'') \end{aligned}$$

$$s_1 = -\alpha f_0 g_0' - \frac{\alpha}{\Pr} g_0''$$

$$s_2 = -\frac{1}{2} f_1 g_1' - \alpha(f_0 g_1' + f_1 g_0') + g_1 f_1' - f_1 g_1' + \alpha(g_1 f_0' - f_1 g_0') - \frac{\alpha}{\Pr} g_1''$$

$$s_3 = -\frac{1}{2}(f_1 g_2' + f_2 g_1') - \alpha(f_0 g_2' + f_1 g_1' + f_2 g_0') + (g_1 f_2' + 2g_2 f_1')$$

$$-(f_1 g_2' + 2f_2 g_1') + \alpha(g_1 f_1' + 2g_2 f_0') - \alpha(f_1 g_1' + 2f_2 g_0') - \frac{\alpha}{\Pr} g_2''$$

$$s_4 = -\frac{1}{2}(f_1 g_3' + f_2 g_2' + f_3 g_1') + (g_1 f_3' + 2g_2 f_2' + 3g_3 f_1')$$

$$-(f_1 g_3' + 2f_2 g_2' + 3f_3 g_1') + \alpha(g_1 f_2' + 2g_2 f_1' + 3g_3 f_0')$$

$$-\alpha(f_1 g_2' + 2f_2 g_1' + 3f_3 g_0') - \frac{\alpha}{\Pr} g_3''$$

$$s_5 = -\frac{\alpha}{\Pr} g_4'' - \frac{1}{2}(f_1 g_4' + f_2 g_3' + f_3 g_2' + f_4 g_1')$$

$$-\alpha(f_0 g_4' + f_1 g_3' + f_2 g_2' + f_3 g_1' + f_4 g_0')$$

$$+(g_1 f_4' + 2g_2 f_3' + 3g_3 f_2' + 4g_4 f_1') - (f_1 g_4' + 2f_2 g_3' + 3f_3 g_2' + 4f_4 g_1')$$

$$+\alpha(g_1 f_3' + 2g_2 f_2' + 3g_3 f_1' + 4g_4 f_0') - \alpha(f_1 g_3' + 2f_2 g_2' + 3f_3 g_1' + 4f_4 g_0')$$

$$s_6 = -\frac{\alpha}{\Pr} g_5'' - \frac{1}{2}(f_1 g_5' + f_2 g_4' + f_3 g_3' + f_4 g_2' + f_5 g_1')$$

$$-\alpha(f_0 g_5' + f_1 g_4' + f_2 g_3' + f_3 g_2' + f_4 g_1' + f_5 g_0')$$

$$+(g_1 f_5' + 2g_2 f_4' + 3g_3 f_3' + 4g_4 f_2' + 5g_5 f_1')$$

$$-(f_1 g_5' + 2f_2 g_4' + 3f_3 g_3' + 4f_4 g_2' + 5f_5 g_1')$$

$$+\alpha(g_1 f_4' + 2g_2 f_3' + 3g_3 f_2' + 4g_4 f_1' + 5g_5 f_0')$$

$$-\alpha(f_1 g_4' + 2f_2 g_3' + 3f_3 g_2' + 4f_4 g_1' + 5f_5 g_0')$$

The zero-order problem defined by Eqs. (6.61)-(6.63) corresponds to constant freestream velocity and subsequent-order problems provide the corrections for its streamwise dependence. A numerical solution of Eqs. (6.64) and (6.65) using the method of superposition has been carried out for $\alpha = \frac{1}{8}$ and $\Pr = 1$. The series for $f''(0)$ and $g'(0)$ can be written from the numerical solutions as

$$f''(0) = 0.3321 + 0.1929\epsilon - 0.05225\epsilon^2 + 0.02371\epsilon^3 - 0.01270\epsilon^4 + 0.008327\epsilon^5 - 0.006115\epsilon^6 \quad (6.66)$$

$$-g'(0) = 0.332060 + 0.0571326\epsilon - 0.0229556\epsilon^2 + 0.013480\epsilon^3 - 0.00894736\epsilon^4 + 0.00634822\epsilon^5 + 0.0042389\epsilon^6 \quad (6.67)$$

6.2.6 Natural Convection from a Nonisothermal Vertical Plate

We now consider an example where the momentum and energy equations are nonsimilar and coupled. Let us consider the problem of natural convection from a nonisothermal vertical plate treated by Kuiken (1969) and Na (1978). The boundary-layer equations of the flow can be written as

$$\frac{\partial u}{\partial x} + \frac{\partial v}{\partial y} = 0$$

$$\rho\left(u\frac{\partial u}{\partial x} + v\frac{\partial u}{\partial y}\right) = \mu\frac{\partial^2 u}{\partial y^2} + \rho g_e \beta(T - T_\infty)$$

$$\rho C\left(u\frac{\partial T}{\partial x} + v\frac{\partial T}{\partial y}\right) = k\frac{\partial^2 T}{\partial y^2}$$

subject to the boundary conditions

$$y = 0 \qquad u = 0 \qquad v = 0 \qquad T = T_w(x)$$

$$y = \infty \qquad u = 0 \qquad T = T_\infty$$

By introducing the transformation

$$\epsilon = \bar{x} \qquad \eta = \frac{\bar{y}}{\bar{x}^{1/4}}[S_w(\bar{x})]^{1/4}$$

$$f(\epsilon, \eta) = \frac{\psi}{\bar{x}^{3/4}}[S_w(\bar{x})]^{-1/4} \qquad g(\epsilon, \eta) = \theta$$

where

$$\bar{x} = \frac{x}{L} \qquad \bar{y} = \frac{y}{L}\sqrt{\mathrm{Re}} \qquad u = \frac{u}{u_c} \qquad v = \frac{v}{u_c}\sqrt{\mathrm{Re}} \qquad \mathrm{Re} = \frac{u_c L}{\nu}$$

$$\theta = \frac{T - T_\infty}{T_w - T_\infty} \qquad u_c = [g_e \beta \cos\alpha (T_r - T) L]^{1/2} \qquad S_w(\bar{x}) = \frac{T_w(x) - T_\infty}{T_r - T_\infty}$$

the boundary-layer equations become

$$f''' + \frac{3 + P(\epsilon)}{4} ff'' - \frac{1}{2}[1 + P(\epsilon)](f')^2 + g = \epsilon\left(f'\frac{\partial f'}{\partial \epsilon} - f''\frac{\partial f}{\partial \epsilon}\right)$$

$$\frac{1}{\mathrm{Pr}} g'' + \frac{3 + P(\epsilon)}{4} fg' - P(\epsilon) f' g = \epsilon\left(f'\frac{\partial g}{\partial \epsilon} - g'\frac{\partial f}{\partial \epsilon}\right)$$

We will now consider the case in which $P(\epsilon) = \epsilon/(1+\epsilon)$, which corresponds to a wall temperature distribution of $T_w \sim 1 + x/L$. For this case, the boundary-layer equations become

$$(1+\epsilon)f''' + \frac{1}{4}(3+4\epsilon)ff'' - \frac{1}{2}(1+2\epsilon)(f')^2 + (1+\epsilon)g$$

$$= \epsilon(1+\epsilon)\ \ f'\frac{\partial g}{\partial \epsilon} - g'\frac{\partial f}{\partial \epsilon} \tag{6.68}$$

$$(1+\epsilon)\frac{1}{\mathrm{Pr}}g'' + \frac{1}{4}(3+4\epsilon)fg' - \epsilon f'g = \epsilon(1+\epsilon)\left(f'\frac{\partial g}{\partial \epsilon} - g'\frac{\partial f}{\partial \epsilon}\right) \tag{6.69}$$

$$f(\epsilon, 0) = f'(\epsilon, 0) = 0 \qquad g(\epsilon, 0) = 1 \tag{6.70}$$

$$f'(\epsilon, \infty) = 0 \qquad g(\epsilon, \infty) = 0 \tag{6.71}$$

where primes denote differentiation with respect to similarity variable η.

Assuming expansions of the form in Eqs. (6.45) and (6.46) and proceeding as usual, the set of equations for the first five terms appear as

$$\epsilon^0: \quad f_0''' + \frac{3}{4}f_0f_0'' - \frac{1}{2}(f_0')^2 + g_0 = 0 \tag{6.72}$$

$$\frac{1}{\mathrm{Pr}}g_0'' + \frac{3}{4}f_0g_0' = 0 \tag{6.73}$$

$$f_0(0) = f_0'(0) = 0 \qquad g_0(0) = 1 \tag{6.74}$$

$$f_0'(\infty) = 0 \qquad g_0(\infty) = 0 \tag{6.75}$$

$$\epsilon^1: \quad f_1''' + \frac{3}{4}f_0f_1'' + \frac{7}{4}f_0''f_1 - 2f_0'f_1' + g_1 = (f_0')^2 - f_0f_0'' - f_0''' - g_0 \tag{6.76}$$

$$\frac{1}{\mathrm{Pr}}g_1'' + \frac{3}{4}f_0g_1' + \frac{7}{4}g_0'f_1 - f_0'g_1 = f_0'g_0 - f_0g_0' - \frac{1}{\mathrm{Pr}}g_0'' \tag{6.77}$$

$$\epsilon^2: \quad f_2''' + \frac{3}{4}f_0f_2'' + \frac{11}{4}f_0''f_2 - 3f_0'f_2' + g_2 = 3f_0'f_1' - 2f_0''f_1 + \frac{3}{2}(f_1')^2$$

$$-\frac{7}{4}f_1f_1'' - f_1''' - f_0f_1'' - g_1 \tag{6.78}$$

$$\frac{1}{\mathrm{Pr}}g_2'' + \frac{3}{4}f_0g_2' + \frac{11}{4}f_2g_0' - 2g_2f_0' = 2f_0'g_1 - 2f_1g_0' + g_1f_1'$$

$$-\frac{7}{4}f_1g_1' - f_0g_1' - f_1g_0' + f_1'g_0 - \frac{1}{\mathrm{Pr}}g_1'' \tag{6.79}$$

$$\epsilon^3:\quad f_3''' + \frac{3}{4} f_0 f_3'' + \frac{3}{4} f_0'' f_3 - f_0' f_3' - 3 f_0' g_3 + 3 g_0' f_3 + g_3$$

$$= -f_2''' - \frac{3}{4} f_1 f_2'' - \frac{3}{4} f_2 f_1'' - f_0 f_2'' - f_1 f_1'' - f_2 f_0'' + f_1' f_2' + 2 f_0' f_2'$$

$$+ (f_1')^2 + f_1' g_1 + 2 f_0' g_2 - f_1 g_1' - 2 f_2 g_0' + g_1 f_2' + 2 g_2 f_1'$$

$$- f_1 g_2' - 2 f_2 g_1' - g_2 \tag{6.80}$$

$$\frac{1}{\mathrm{Pr}} g_3''' + \frac{3}{4} f_0 g_3' + \frac{3}{4} g_0' f_3 - 3 f_0' g_3 + 3 g_0' f_3$$

$$= -2 f_1 g_2' - 3 f_2 g_1' - f_0 g_2' - 2 f_1 g_1' - 3 f_2 g_0' + f_0' g_2 + 2 f_1' g_1$$

$$+ f_2' g_0 + 2 f_0' g_2 + f_2' g_1 + 2 f_1' g_2 - \frac{1}{\mathrm{Pr}} g_2'' \tag{6.81}$$

$$\epsilon^4:\quad f_4''' + \frac{3}{4} f_0 f_4'' - 5 f_0' f_4' + \frac{19}{4} f_0'' f_4 + g_4$$

$$= -f_0 f_3'' - 2 f_1 f_2'' - 3 f_1'' f_2 - 4 f_0'' f_3 - \frac{7}{4} f_1 f_3'' - 3 f_2 f_2'' - 4 f_1'' f_3$$

$$+ 5 f_1' f_3' + \frac{5}{2} (f_2')^2 + 5 f_0' f_3' + 4 f_1' f_2' - f_3''' - g_3 \tag{6.82}$$

$$\frac{1}{\mathrm{Pr}} g_4'' + \frac{3}{4} f_0 g_4' + \frac{19}{4} g_0' f_4 - 4 f_0' g_4 = -2 f_1 g_3' - 3 f_2 g_2' - \frac{15}{4} f_3 g_1'$$

$$- f_0 g_3' - 2 f_1 g_2' - 3 f_2 g_1' - 4 f_3 g_0' + 4 f_0' g_3 + 3 f_1' g_2 + 2 f_2' g_1$$

$$+ f_3' g_0 + f_3' g_1 + 2 f_2' g_2 + 3 f_1' g_3 - \frac{1}{\mathrm{Pr}} g_3'' \tag{6.83}$$

The numerical solutions up to $O(\epsilon^2)$ have been obtained by Kuiken (1969); we have extended them up to $O(\epsilon^4)$. Confining attention to the series for $g'(0)$ which is related to the heat transfer rate

$$\frac{\mathrm{Nu}_x}{(\mathrm{Gr}_x)^{1/4}} = -g'(0) \tag{6.84}$$

we have for $\mathrm{Pr} = 0.73$,

$$-g'(0) = 0.358600 + 0.207576\epsilon - 0.230733\epsilon^2 + 0.265562\epsilon^3$$

$$- 0.267053\epsilon^4 \tag{6.85}$$

6.2.7 Miscellaneous Examples

In the foregoing section we have given examples with full details about the derivation of the sequence of perturbation equations and quoted the final results. The reader should by now have sufficient insight into the mechanics of extending the series. Now we present a number of examples of extended perturbation series, leaving out the details to the appropriate references.

First, Beckett (1981) considers the inward freezing of a cylinder and extends the series for the freezing front position E in time ϵ to 31 terms. For $\beta = 0.1$, his result is

$$\begin{aligned} E &= 2.5139\epsilon^{1/2} + 0.5130\epsilon + 0.6540\epsilon^{3/2} + 1.2779\epsilon^2 + 2.2475\epsilon^{5/2} \\ &+ 4.8757\epsilon^3 + 11.183\epsilon^{7/2} + 26.677\epsilon^4 + 65.527\epsilon^{9/2} + 164.64\epsilon^5 \\ &+ 421.22\epsilon^{11/2} + 1093.6\epsilon^6 + 2874.5\epsilon^{13/2} + 7643.2\epsilon^7 + \cdots \\ &+ 8.651 \times 10^6\epsilon^{21/2} + \cdots + 2.759 \times 10^{11}\epsilon^{31/2} \end{aligned} \tag{6.86}$$

Second, Van Dyke (1977) uses Morton's (1960) results to derive a 20-term series for the volume flux ratio Q/Q_0 in terms of Rayleigh number ϵ for mixed convection in a vertical pipe

$$\begin{aligned} \frac{Q}{Q_0} &= 1 - \frac{11}{12}\frac{\epsilon}{32} + \frac{473}{540}\left(\frac{\epsilon}{32}\right)^2 - \frac{101{,}369}{120{,}960}\left(\frac{\epsilon}{32}\right)^3 + \frac{65{,}467{,}219}{81{,}648{,}000}\left(\frac{\epsilon}{32}\right)^4 \\ &- \cdots + 0.432014\left(\frac{\epsilon}{32}\right)^{18} - 0.143346\left(\frac{\epsilon}{32}\right)^{19} \end{aligned} \tag{6.87}$$

Third, we have calculated the first five terms of nonsimilar boundary-layer equations for laminar natural convection over a frustum of a cone. For $\mathrm{Pr} = 1$, the series for $\mathrm{Nu}_x/\mathrm{Gr}_x^{1/4}$ in terms of transverse curvature parameter ϵ is

$$\begin{aligned} \frac{\mathrm{Nu}_x}{\mathrm{Gr}_x^{1/4}} &= -g'(0) = 0.401100 + 0.0894649\epsilon - 0.0716556\epsilon^2 \\ &+ 0.0627372\epsilon^3 - 0.0498746\epsilon^4 \end{aligned} \tag{6.88}$$

Fourth, Afzal (1981) deals with mixed convection in a two-dimensional buoyant plume and calculates an 11-term series for the velocity F' and temperature H. In terms of $\epsilon = \mathrm{Gr}_x/\mathrm{Re}_x^{5/2}$, the series for $\mathrm{Pr} = 0.72$ are

$$F'(0) = \sum_{n=0}^{\infty} a_n(\pm\epsilon)^n \tag{6.89}$$

$$H(0) = \sum_{n=0}^{\infty} b_n(\pm\epsilon)^n \tag{6.90}$$

where a_n and b_n are given in Table 6.2. The plus sign corresponds to assisting flow and the minus sign corresponds to opposing flow.

Table 6.2 Coefficients a_n and b_n of series in Eqs. (6.89) and (6.90)

n	a_n	b_n
0	1.0	0.239365
1	0.258975	-0.125705×10^{-1}
2	-0.51126×10^{-1}	0.301922×10^{-2}
3	0.174952×10^{-1}	-0.101311×10^{-2}
4	-0.722383×10^{-2}	0.397441×10^{-3}
5	0.32893×10^{-2}	-0.170991×10^{-3}
6	-0.159151×10^{-2}	0.782146×10^{-4}
7	0.802681×10^{-3}	-0.373848×10^{-4}
8	-0.417275×10^{-3}	0.184713×10^{-4}
9	0.221987×10^{-3}	-0.936619×10^{-5}
10	-0.120264×10^{-3}	0.484946×10^{-5}

Fifth, Anderson and Reynolds (1970) have extended Morton's (1959) series for the ratio of Nusselt numbers N/N_0 for mixed convection in a horizontal pipe. Their result is

$$\frac{N}{N_0} = 1 + 0.04891\left(\frac{\epsilon}{1152}\right)^2 + 0.15652\left(\frac{\epsilon}{1152}\right)^4 + 0.0073903\left(\frac{\epsilon}{1152}\right)^6 \tag{6.91}$$

Sixth, Aziz (1981) has extended to 10 terms the series for midplane temperature in a slab with exponential heat generation. It reads

$$\begin{aligned}\theta(0) = \epsilon(0.5000 &+ 0.208333\epsilon + 0.130555\epsilon^2 + 0.097070\epsilon^3 \\ &+ 0.089537\epsilon^4 + 0.085060\epsilon^5 + 0.082508\epsilon^6 + 0.083333\epsilon^7 \\ &+ 0.085833\epsilon^8 + 0.088838\epsilon^9)\end{aligned} \tag{6.92}$$

Seventh, the base heat flux during the transient response of an infinitely long convecting fin is given by (Aziz and Na, 1980)

$$Q\sqrt{\pi\tau} = e^{(-1/4)\epsilon} + \frac{\sqrt{\pi}}{2}\,\epsilon^{1/2}\,\mathrm{erf}\left(\frac{1}{2}\,\epsilon^{1/2}\right) \tag{6.93}$$

which can be extended to give

$$\begin{aligned}Q\sqrt{\pi\tau} = 1 &+ 0.25\epsilon - 0.0104\epsilon^2 + 5.2083 \times 10^{-4}\epsilon^3 - 2.3251 \times 10^{-5}\epsilon^4 \\ &+ 9.0420 \times 10^{-7}\epsilon^5 - 3.0826 \times 10^{-8}\epsilon^6 + 9.3155 \times 10^{-10}\epsilon^7 \\ &- 2.5229 \times 10^{-11}\epsilon^8 + \cdots + 8.0828 \times 10^{-31}\epsilon^{19} \\ &- 9.5857 \times 10^{-33}\epsilon^{20} + \cdots\end{aligned} \tag{6.94}$$

Table 6.3 Coefficients d_n of series in Eq. (6.95)

n	d_n	n	d_n	n	d_n
0	5.64190×10^{-1}	13	4.40537×10^{-7}	26	-7.59019×10^{-8}
1	-2.07538×10^{-2}	14	4.34415×10^{-7}	27	-3.02532×10^{-7}
2	-2.91608×10^{-3}	15	4.06692×10^{-7}	28	-6.80280×10^{-7}
3	-7.05065×10^{-4}	16	3.79290×10^{-7}	29	-1.28919×10^{-6}
4	-2.17511×10^{-4}	17	3.51029×10^{-7}	30	-2.29744×10^{-6}
5	-7.54497×10^{-5}	18	3.27831×10^{-7}	31	-3.91132×10^{-6}
6	-2.81048×10^{-5}	19	3.06039×10^{-7}	32	-6.52786×10^{-6}
7	-1.08140×10^{-5}	20	2.86578×10^{-7}	33	-1.05735×10^{-5}
8	-4.00950×10^{-6}	21	2.65487×10^{-7}	34	-1.66536×10^{-5}
9	-1.31237×10^{-6}	22	2.40081×10^{-7}	35	-2.46785×10^{-5}
10	-1.84130×10^{-7}	23	2.04763×10^{-7}	36	-3.23612×10^{-5}
11	2.43706×10^{-7}	24	1.50336×10^{-7}		
12	4.07153×10^{-7}	25	6.51076×10^{-8}		

Finally, Van Dyke (1975b) gives a 37-term series for heat transfer rate $-\theta'(0)$ in impulsive heating of the boundary-layer on a flat plate

$$-\theta'(0) = \sum_{n=0}^{\infty} \epsilon^{(3n-1)/2} d_n \tag{6.95}$$

where d_n are tabulated in Table 6.3.

6.3 ANALYSIS OF SERIES

The previous section dealt with the extension of the perturbation series. Once this is accomplished, the next and the most crucial step is to explore the analytic structure of the solution. The analysis will be aimed at identifying the location and nature of singularities characterizing the series.

6.3.1 Pattern of Signs

We begin by examining the final pattern of signs appearing in the extended series. Since we have only a finite number of terms available, it is impossible to predict with certainty what pattern of signs will ultimately emerge. However, one may reasonably hope that the pattern exhibited by a finite number of terms will continue to prevail.

The signs can be all positive as in Eq. (6.86), Eq. (6.90) for opposing flow, and Eqs. (6.91) and (6.92); all negative as in Eq. (6.89) for opposing flow; and alternating as in Eqs. (6.17), (6.31), (6.40), (6.55), (6.56), (6.66),

(6.67), (6.85), (6.87), and (6.80). Eqs. (6.89) and (6.90) for assisting flow, and Eq. (6.94). The behavior of Eq. (6.95) is rather atypical but hopefully an all negative sign prevails ultimately. On rare occasions, we may also encounter series with random patterns of signs (Shih, 1981).

The pattern of signs determines the direction of the dominant singularity. If the signs are fixed, this singularity lies on the positive axis, but if the signs alternate, it lies on the negative axis. With random signs, the most likely possibility is that singularities occur as complex conjugate pairs.

Most often the perturbation quantity ϵ takes positive values; it can sometimes be negative or even imaginary. For example, in Eq. (6.18) it can assume both positive and negative values; positive if thermal conductivity increases with temperature, and negative if thermal conductivity decreases with temperature. For the sake of generality, we treat ϵ as a complex quantity to scan the entire complex plane for singularities.

6.3.2 Significance of Nearest Singularity

When the nearest singularity lies on the negative axis, it usually carries no physical significance. Thus in the series in Eqs. (6.17), (6.40), and (6.55), ϵ represents the radiation-conduction parameter, transverse curvature parameter, and wedge angle, respectively; a negative value of ϵ is meaningless. On the other hand, a singularity appearing on the positive axis often has a physical interpretation. One such example is the series in Eq. (6.92) for the mid-plane temperature in a heat generating slab. Here, as will be seen later, the singularity represents the critical heat generation beyond which a steady-state solution does not exist. Similarly, the singularity associated with the series in Eq. (6.86) indicates the completion of the freezing process in a cylinder. In some cases, the positive axis singularity points to the limit of validity of the mathematical model itself. This situation is encountered in Eq. (1.55) where, for $\epsilon = 1$, the term $(1 - \epsilon f')$ vanishes at the edge of the boundary-layer. As a consequence, the unsteady term is lost and the partial differential equation reduces prematurely to an ordinary differential equation (Cebeci and Bard, 1973; Ingham, 1977), which is actually applicable at $\epsilon = \infty$. Another possibility with the positive axis singularity is that it may not be real, but simply an indication that the function is double-valued. This situation arises with the series in Eq. (6.89) where the dual solutions correspond to the forward and reverse flows in the plume.

6.3.3 Location and Nature of Singularities

We now turn to the crucial problem of determining the location and nature of nearest singularity. If an infinite number of terms of a power series $f(\epsilon) =$

$\Sigma_{n=0}^{\infty} c_n \epsilon^n$ are known, the radius of convergence ϵ_0 can be calculated using D'Alembert's ratio limit

$$\epsilon_0 = \lim_{n \to \infty} \left| \frac{c_{n-1}}{c_n} \right| \tag{6.96}$$

However, most perturbation series have a finite number of terms. The best we can do is to use the available information to extrapolate to $n = \infty$ and estimate ϵ_0 approximately. Such an attempt for the series in Eq. (6.17) is shown in Fig. 6.1 where $|c_{n-1}/c_n|$ is plotted against n. It is clear that extrapolation to $n = \infty$ is difficult. To overcome this difficulty, Domb and Sykes (1957) suggested a plot of inverse ratio $|c_n/c_{n-1}|$ versus $1/n$ (called Domb-Sykes plot) which can then be extrapolated to $1/n = 0$ to estimate $1/\epsilon_0$. Figure 6.2 shows such a plot for the series in Eq. (6.17). As n increases, the plot becomes almost linear. Assuming the linear pattern prevails, one can extrapolate lineary to $1/n = 0$ and the intercept gives $1/\epsilon_0$. This assumption should be valid if the nearest singularity is such that

$$f(\epsilon) = \sum_{n=0}^{\infty} c_n \epsilon^n = \text{constant} \times \begin{cases} (\epsilon_0 \pm \epsilon)^\alpha & \alpha \neq 0, 1, 2 \quad (6.97a) \\ (\epsilon_0 \pm \epsilon)^\alpha \log(\epsilon_0 \pm \epsilon) & \alpha = 0, 1, 2 \quad (6.97b) \end{cases}$$

Then for large n the ratio c_n/c_{n-1} is exactly linear in $1/n$, the relationship being

$$\frac{c_n}{c_{n-1}} = \pm \frac{1}{\epsilon_0} \left(1 - \frac{1 + \alpha}{n} \right) \tag{6.98}$$

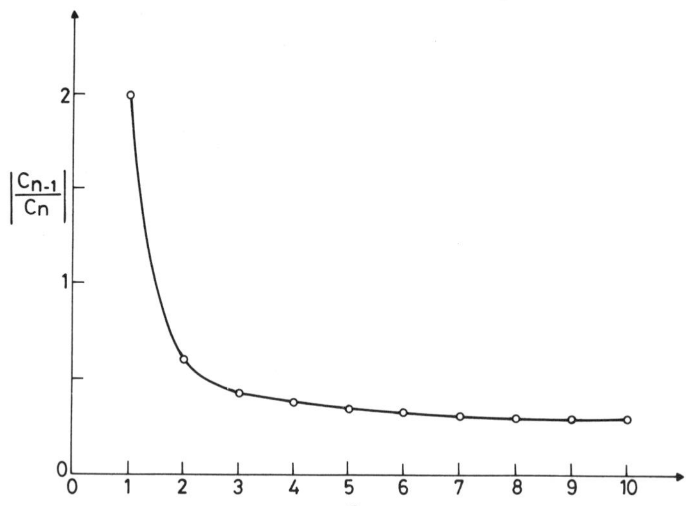

Figure 6.1 D'Alembert ratio test for tip temperature in a radiating fin for the series in Eq. (6.17).

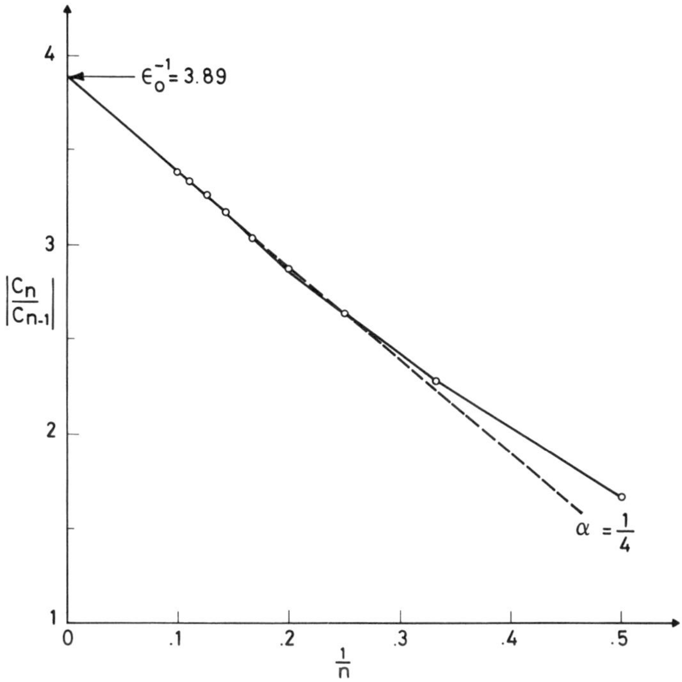

Figure 6.2 Domb-Sykes plot for tip temperature in a radiating fin for the series in Eq. (6.17).

Thus the intercept for $1/n = 0$ gives $1/\epsilon_0$ and the slope permits evaluation of exponent α. For the case using Eq. (6.97*b*), this relation holds only for $n \geqslant 2 + \alpha$. If the function is more complicated, but the nearest singularity has a leading term of the form of Eq. (6.97), the ratio c_n/c_{n-1} will behave asymptotically like Eq. (6.98) for large n.

While extrapolating one must adjust the slope slightly so that the exponent α is an integer or a simple fraction. Sometimes the correct estimate of ϵ_0 and α is also guided by analyses other than perturbation.

6.3.4 Examples of Domb-Sykes Plot

A total of eight series have been chosen from Section 6.2 to draw the Domb-Sykes plots. These appear in Figs. 6.2–6.9 and exhibit different characteristics. In Fig. 6.2–6.5, the linear trend emerges with only a few terms and extrapolation is straightforward. The only care that was exercised in drawing the graphs was to adjust the slope slightly to give simple values of exponent α. However, in Fig. 6.6 we have drawn two curves, one for the series in Eq. (6.89) and the other for the series in Eq. (6.90). The curves are slightly

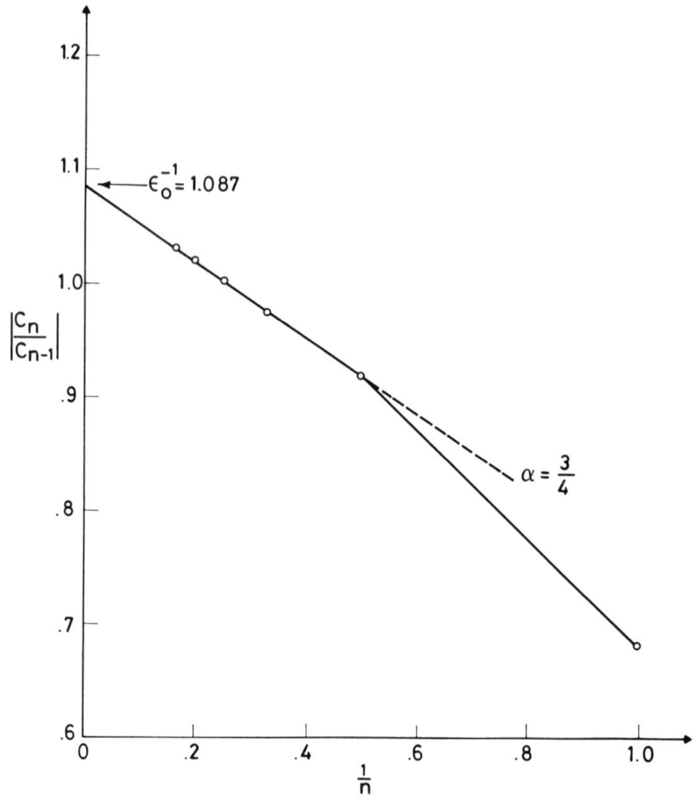

Figure 6.3 Domb-Sykes plot for wall heat flux in semi-infinite medium for the series in Eq. (6.31).

displaced and, following Afzal (1981), we have drawn the extrapolation line so that it falls between the two curves. In Fig. 6.7, the plot is well-behaved and approaches the origin, indicating an infinite radius of convergence. Figure 6.8 shows how extrapolation with a few terms can be fallacious. With just five terms, a value of $\alpha = \frac{1}{3}$ may be deduced, but with eight terms, a value of $\alpha = -1$ seems more appropriate. In rare instances, the plot may exhibit random behavior as typified by Fig. 6.9. Any hope of extrapolation is lost and one should seek a different route to estimate the radius of convergence. For this particular series, Van Dyke (1975b) applied the root test and concluded that the radius of convergence is unity ($\epsilon_0 = 1$).

6.4 IMPROVEMENT OF SERIES

It must be admitted that our exploration of the analytic structure of the series has been brief and rather crude. But this need not cause us to despair.

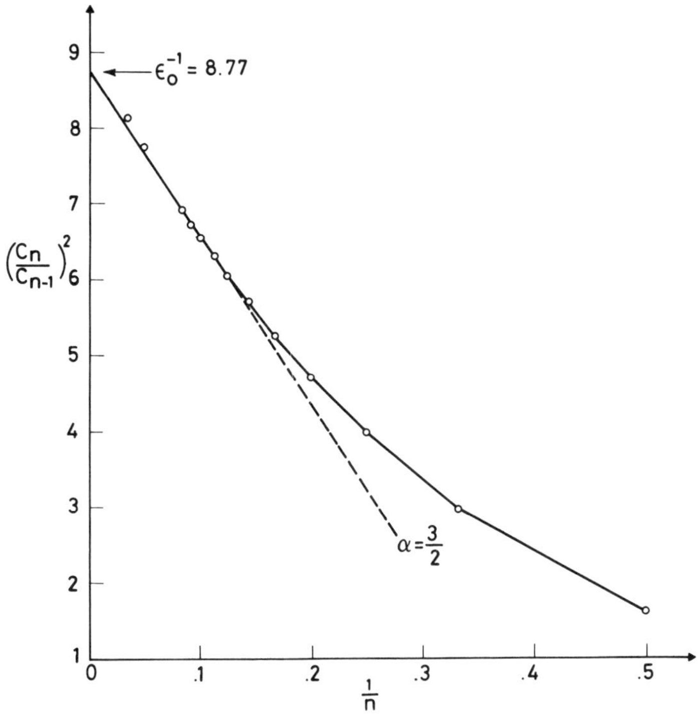

Figure 6.4 Domb-Sykes plot for freezing of a cylinder for the series in Eq. (6.86).

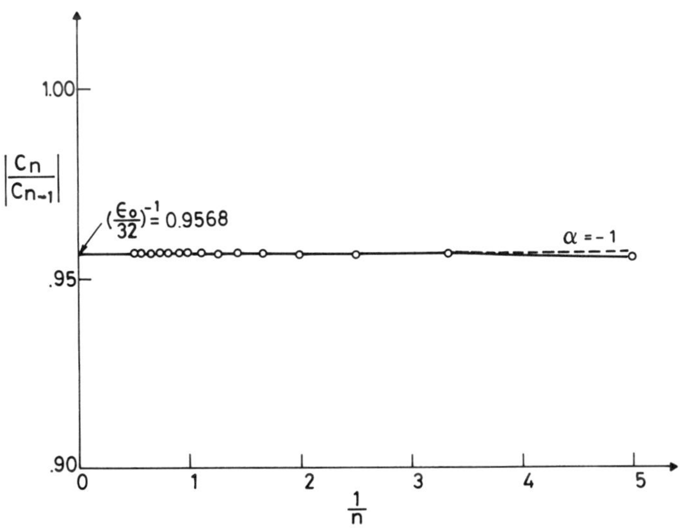

Figure 6.5 Domb-Sykes plot for the volume flux ratio in mixed convection in a vertical pipe for the series in Eq. (6.87).

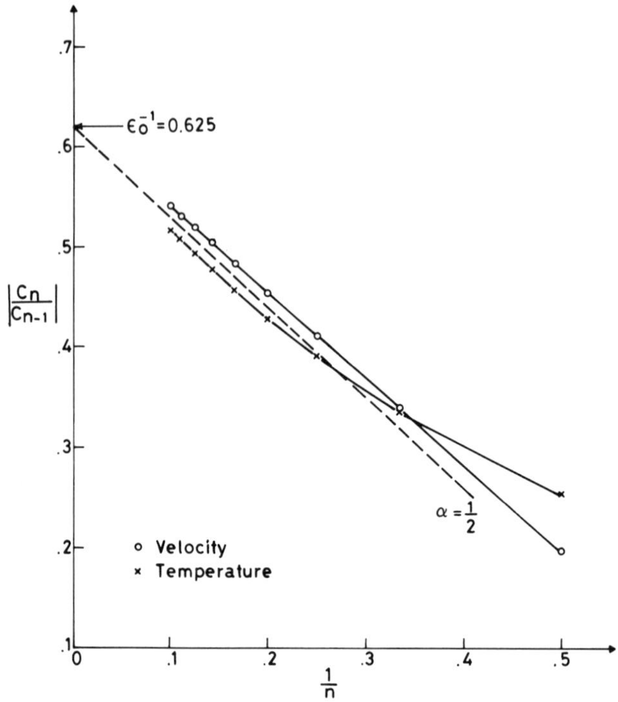

Figure 6.6 Domb-Sykes plot for velocity and temperature for the series in Eqs. (6.89) and (6.90).

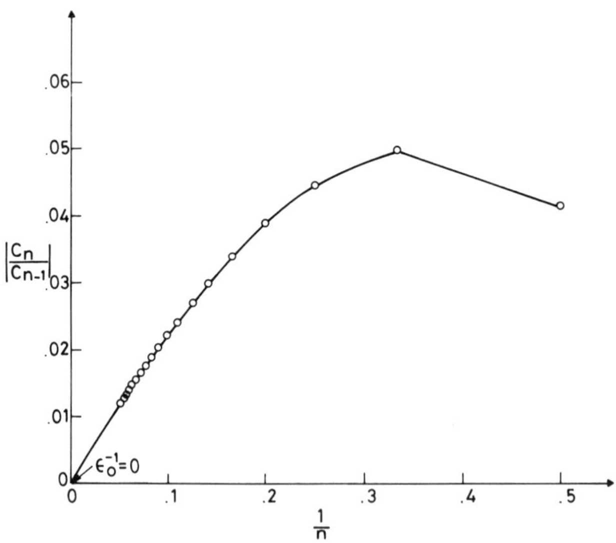

Figure 6.7 Domb-Sykes plot for base heat flux in an infinitely long convecting fin for the series in Eq. (6.94).

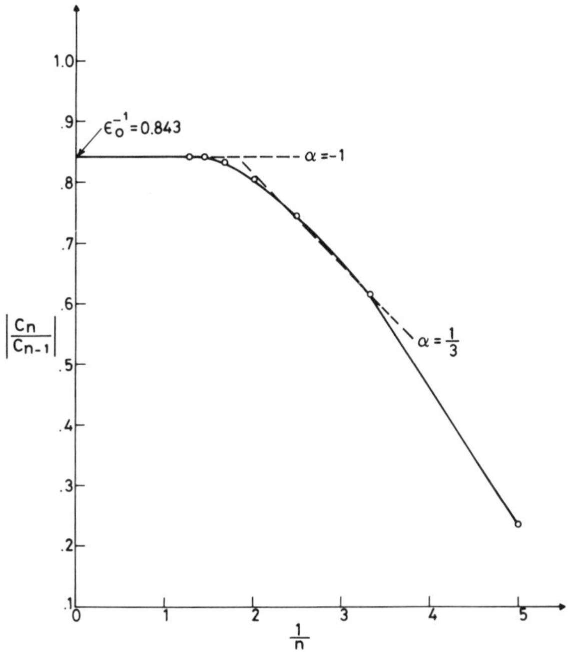

Figure 6.8 Domb-Sykes plot for friction in longitudinal flow over a cylinder for the series in Eq. (6.40).

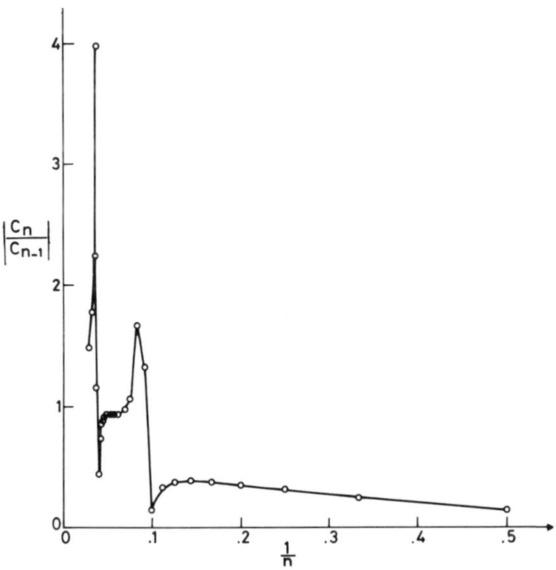

Figure 6.9 Domb-Sykes plot for heat flux in an impulsively heated flat plate for the series in Eq. (6.95).

We proceed further and make use of our knowledge about the leading singularity to improve the series. We describe here a number of improvement devices and illustrate their use. The choice of a particular device depends on the direction, distance, and nature of the singularity as revealed by the Domb-Sykes plot. Often the problem is such that more than one device is applicable, and it is always enlightening to try different possibilities. Also, if a single technique is not adequate, one should consider using two or more of them in combination.

6.4.1 Euler Transformation

When a perturbation series contains alternating signs, the singularity lies on the negative axis, and carries no physical significance. In this case, the simplest device to use is an Euler transformation based on the estimate of ϵ_0. With this transformation the singularity is mapped to infinity. The advantage of this device is that the exact nature of the singularity need not be known.

Let the perturbation series be

$$f = \sum_{n=0}^{\infty} \epsilon^n a_n \tag{6.99}$$

and the nearest singularity be located at $\epsilon = \epsilon_0$ (estimated from Domb-Sykes plot). The transformation envisages using a new variable ϵ^* such that

$$\epsilon^* = \frac{\epsilon}{\epsilon + \epsilon_0} \qquad \text{or } \epsilon = \frac{\epsilon_0 \epsilon^*}{1 - \epsilon^*} \tag{6.100}$$

Substituting Eq. (6.100) into Eq. (6.99), we get the Eulerized series

$$f = \sum_{n=0}^{\infty} b_n \epsilon^{*n} \tag{6.101}$$

where the coefficients b_n are

$$b_0 = a_0 \tag{6.102}$$

$$b_n = \sum_{j=1}^{n} \frac{(n-1)!}{(n-j)!\,(j-1)!} a_j \epsilon_0^j \tag{6.103}$$

For example, the six terms of Eq. (6.101) can be written as

$$\begin{aligned} f = {} & a_0 + (a_1 \epsilon_0)\epsilon^* + (a_1 \epsilon_0 + a_2 \epsilon_0^2)\epsilon^{*2} + (a_1 \epsilon_0 + 2a_2 \epsilon_0^2 + a_3 \epsilon_0^3)\epsilon^{*3} \\ & + (a_1 \epsilon_0 + 3a_2 \epsilon_0^2 + 3a_3 \epsilon_0^3 + a_4 \epsilon_0^4)\epsilon^{*4} \\ & + (a_1 \epsilon_0 + 4a_2 \epsilon_0^2 + 6a_3 \epsilon_0^3 + 4a_4 \epsilon_0^4 + a_5 \epsilon_0^5)\epsilon^{*5} \end{aligned} \tag{6.104}$$

To illustrate the above procedure, we apply it to two series. First, consider the series in Eq. (6.17). Figure 6.2 gives $\epsilon_0 = -0.25707$. The transformed series becomes

$$\theta(0) = 1 - 0.128535\epsilon^* - 0.073464\epsilon^{*2} - 0.050766\epsilon^{*3} - 0.038416\epsilon^{*4} - 0.030690\epsilon^{*5} - 0.025443\epsilon^{*6} - 0.021661\epsilon^{*7} - 0.019002\epsilon^{*8} - 0.017254\epsilon^{*9} - 0.015804\epsilon^{*10} \tag{6.105}$$

The fixed pattern of negative signs in Eq. (6.105) shows that the nearest singularity now lies on the positive axis. It is interesting to note that the coefficients have been greatly reduced. The radius of convergence is now $\epsilon^* = 1$ which means that the series now converges right up to $\epsilon = \infty$. This is a significant improvement over the original series which is computationally useless for higher values of ϵ. However, it must be emphasized that the convergence of the transformed series becomes slower as ϵ increases. With finite terms, the series may fail to give the correct limit as $\epsilon \to \infty$. For example, at $\epsilon^* = 1$ or $\epsilon = \infty$, the eleven terms of Eq. (6.105) give the erroneous value of $\theta(0) = 0.5790$ instead of the true value of zero dictated by physical consideration. However, this fact is of little consequence because $\epsilon = 4$ is the upper limit of radiating fin usefulness in practice. To demonstrate the accuracy of Eq. (6.105) we use it to calculate the fin efficiency η

$$\eta = 2\left[\frac{1-\theta(0)}{10\epsilon}\right]^{1/2} \tag{6.106}$$

A plot of η versus ϵ obtained from Eq. (6.106) is compared with the numerical solution (Sparrow and Cess, 1978) in Fig. 6.10 for the range $\epsilon = 0$ to 4. The remarkable agreement between the two results vouches for the success of Euler transformation.

As another example, we consider the series in Eq. (6.31) for the wall heat flux in a semi-infinite medium. From Fig. 6.3, $\epsilon_0 = (1.087)^{-1} = 0.9199$. The transformed series follows as

$$-\theta'(0) = \frac{2}{\sqrt{\pi}}(1 - 0.62701\epsilon^* - 0.096513\epsilon^{*2} - 0.041280\epsilon^{*3} - 0.02329\epsilon^{*4} - 0.015470\epsilon^{*5} - 0.012367\epsilon^{*6}) \tag{6.107}$$

where again a fixed pattern of signs has emerged indicating that the nearest singularity now appears on the positive axis. The radius of convergence of $\epsilon^* = 1$ implies that the convergence has been extended to $\epsilon = \infty$. In Table 6.4 values of $-\theta'(0)$ from the original series in Eq. (6.31) and the Eulerized series in Eq. (6.107) are compared to the numerical solution. The accuracy of the original series deteriorates rapidly beyond $\epsilon = 0.5$ and produces wild answers

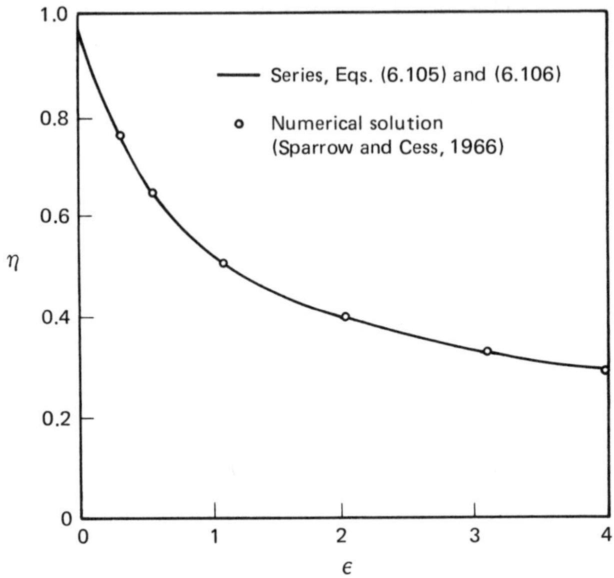

Figure 6.10 Efficiency of a radiating fin.

as ϵ increases but the transformed series in Eq. (6.107) is remarkably accurate even at $\epsilon = 4$. It is improbable that the thermal conductivity variation in any engineering application would be so drastic as to exceed the limit $\epsilon = 4$. Thus, with Euler transformation, we have succeeded in developing a highly accurate analytical solution to a classical nonlinear heat conduction problem.

6.4.2 Extraction of Singularity

The extraction of the nearest singularity is another useful technique to accelerate convergence. It is applicable irrespective of whether the singularity

Table 6.4 Values of $-\theta'(0)$

ϵ	Original series, Eq. (6.31)	Improved series, Eq. (6.107)	Numerical solution
0	1.1284	1.1284	1.1284
0.5	0.8670	0.8632	0.8631
1.0	1.0892	0.6986	0.6982
2.0	31.8473	0.5678	0.5642
3.0	400.2361	0.4858	0.4768
4.0	2120.1477	0.4344	0.4275

lies on the positive or negative axis. If the exponent α is negative, the singularity can be either multiplicative or additive, and extracted either way. However, it is simpler to extract it multiplicatively to avoid estimating its amplitude. On the other hand, if α is positive, it is advisable to estimate the amplitude of the singularity and extract it by subtraction. Here, we discuss the former situation and try it on two series.

Assuming that the singularity lies on the positive axis at $\epsilon = \epsilon_0$ and the exponent α is negative, we can extract it by writing

$$f = \sum_{n=0}^{\infty} \epsilon^n a_n = (\epsilon_0 - \epsilon)^{-\alpha} \sum_{n=0}^{\infty} b_n \epsilon^n \tag{6.108}$$

where the coefficients b_n can be readily determined by multiplying the original series $\Sigma_{n=0}^{\infty} \epsilon^n a_n$ by $(\epsilon_0 - \epsilon)^{\alpha}$. For example, the first four coefficients are

$$b_0 = a_0 \epsilon_0^{\alpha} \tag{6.109a}$$

$$b_1 = a_1 \epsilon_0^{\alpha} - a_0 \alpha \epsilon_0^{\alpha-1} \tag{6.109b}$$

$$b_2 = a_2 \epsilon_0^{\alpha} - a_1 \alpha \epsilon_0^{\alpha-1} + \frac{a_0 \alpha(\alpha-1)}{2!} \epsilon_0^{\alpha-2} \tag{6.109c}$$

$$b_3 = a_3 \epsilon_0^{\alpha} - a_2 \alpha \epsilon_0^{\alpha-1} + \frac{a_1 \alpha(\alpha-1)}{2!} \epsilon_0^{\alpha-2} - \frac{a_0 \alpha(\alpha-1)(\alpha-2)}{3!} \epsilon_0^{\alpha-3} \tag{6.109d}$$

Obviously, if the singularity lies on the negative axis, we only need to replace ϵ by $-\epsilon$. Similarly, if the exponent α is positive, a change in sign of α is all that is needed.

For illustration, let us choose the series in Eq. (6.92). The Domb-Sykes plot is shown in Fig. 6.11 from which $\epsilon_0 = 0.88$ and $\alpha = -\frac{1}{10}$. Extracting the singularity, we have

$$\begin{aligned} \theta(0) = \epsilon(0.88 - \epsilon)^{-0.1}(&0.493650 + 0.149559\epsilon + 0.076840\epsilon^2 \\ &+ 0.099698\epsilon^3 + \cdots) \end{aligned} \tag{6.110}$$

Although the extraction of singularity has not reduced the coefficients significantly, the convergence in the neighborhood of the critical point has improved. For example, at $\epsilon = 0.8$, the numerical solution gives $\theta(0) = 0.7517$. The four terms of the original series give 0.6399 which is about 15 percent in error whereas the four terms of the improved series give 0.7348 which is only 2 percent in error. To obtain an accuracy of 2 percent with the original series at least 9 terms are needed.

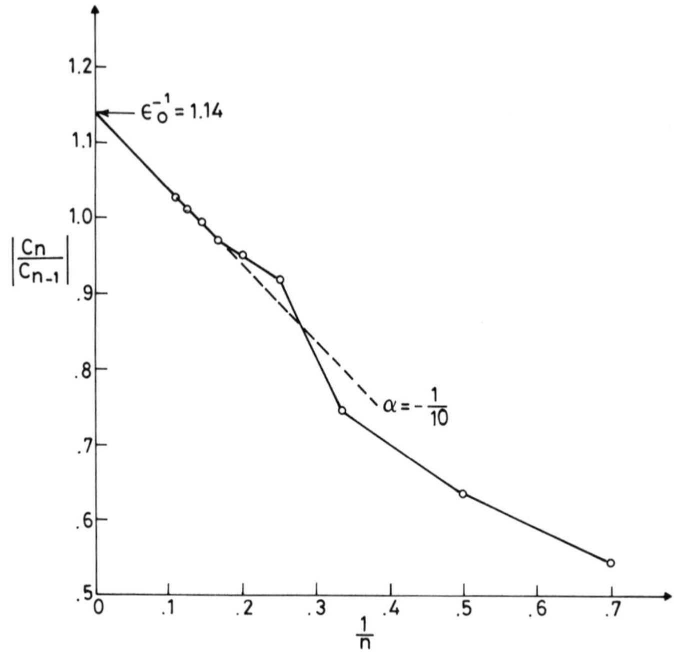

Figure 6.11 Domb-Sykes plot for mid-plane temperature in a slab with exponential heat generation for the series in Eq. (6.92).

Next, let us try to improve the series in Eq. (6.87). From its Domb-Sykes plot in Fig. 6.5, we estimate $\epsilon_0/32 = 1.045517$. Extracting the pole, we obtain

$$\frac{Q}{Q_0} = \left(1.04517 + \frac{\epsilon}{32}\right)^{-1}\left[1.04517 + 0.0419275\,\frac{\epsilon}{32} - 0.001752\left(\frac{\epsilon}{32}\right)^2 + 0.0000344\left(\frac{\epsilon}{32}\right)^3 - \cdots\right] \tag{6.111}$$

In this case the improvement is quite remarkable. While the convergence of the original series in Eq. (6.87) is limited to $\epsilon \simeq 33$, this limit is extended to $\epsilon \simeq 930$ for the improved series in Eq. (6.111). For example, at $\epsilon = 1000$, the original series gives a wild answer, but the improved series gives $Q/Q_0 = 0.069905$ compared to the exact value of 0.063662. Figure 6.12 compares the series in Eqs. (6.87) and (6.111) with the exact solution.

6.4.3 Reversion of Series

It has been remarked previously that a singularity appearing on the positive axis is often real and represents a physical limit. However, this is not so in the exceptional case of an isolated square root ($\alpha = \frac{1}{2}$) singularity. The singularity has no physical interpretation but simply indicates that the function is double-valued. A simple device to eliminate the singularity is to reverse the series so as to interchange the roles of the independent and dependent variables.

Let the original series be

$$f - a_0 = \sum_{n=1}^{\infty} a_n \epsilon^n \tag{6.112}$$

then the reversed series can be written as

$$\epsilon = \sum_{n=1}^{\infty} b_n (f - a_0)^n \tag{6.113}$$

where the first four coefficients b_n, for example, are

$$b_1 = \frac{1}{a_1} \qquad b_2 = -\frac{a_2}{a_1^3} \qquad b_3 = \frac{2a_2^2 - a_1 a_3}{a_1^5} \qquad b_4 = \frac{5a_1 a_2 a_3 - a_1^2 a_4 - 5a_2^3}{a_1^7} \tag{6.114}$$

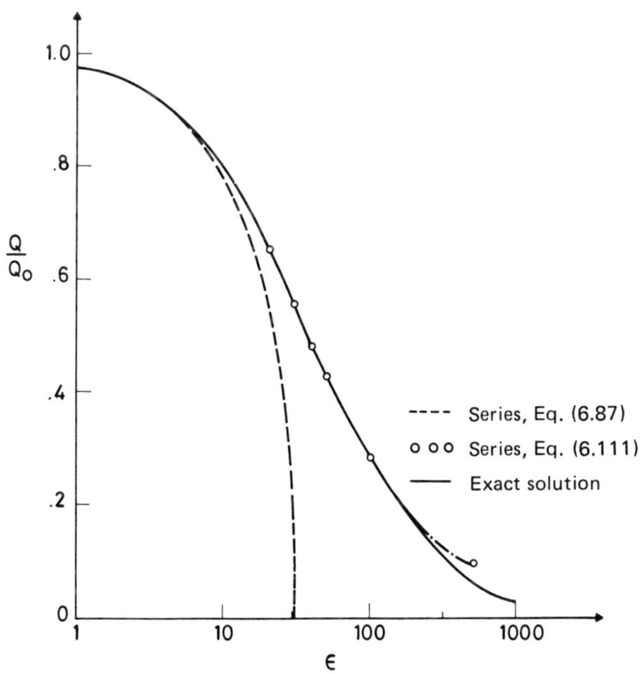

Figure 6.12 Comparison of the series in Eqs. (6.87) and (6.111) and the exact solution.

To illustrate the above, we consider the problem of mixed convection in a two-dimensional buoyant plume in opposing flow. The series for the velocity follows from Eq. (6.89) with the coefficients taken from Table 6.2. Truncating the expansion at ϵ^4, we have

$$F'(0) - 1 = -0.258975\epsilon - 0.051126\epsilon^2 - 0.0174952\epsilon^3 - 0.00722383\epsilon^4 \tag{6.115}$$

Using Eqs. (6.113) and (6.114), the reversed series become

$$\epsilon = -3.861376[F'(0)-1] - 2.943529[F'(0)-1]^2 - 0.598262[F'(0)-1]^3 + 0.070922[F'(0)-1]^4 \tag{6.116}$$

The results from Eqs. (6.115) and (6.116) are plotted in Fig. 6.13 and compared with the solution reported by Afzal (1981). Whereas the series in Eq. (6.115) starts to deviate beyond $\epsilon = 0.9$, the series in Eq. (6.116) follows Afzal's solution right up to $\epsilon = 1.6$, the point of flow reversal. Despite its simplicity the series in Eq. (6.116) is as accurate as the rather involved solution derived by Afzal by subtracting the singularity and calculating ten additional terms to complete the series.

6.4.4 Shanks Transformation

In a pioneering paper Shanks (1955) introduced a family of four nonlinear transformations to accelerate the convergence of slowly convergent and

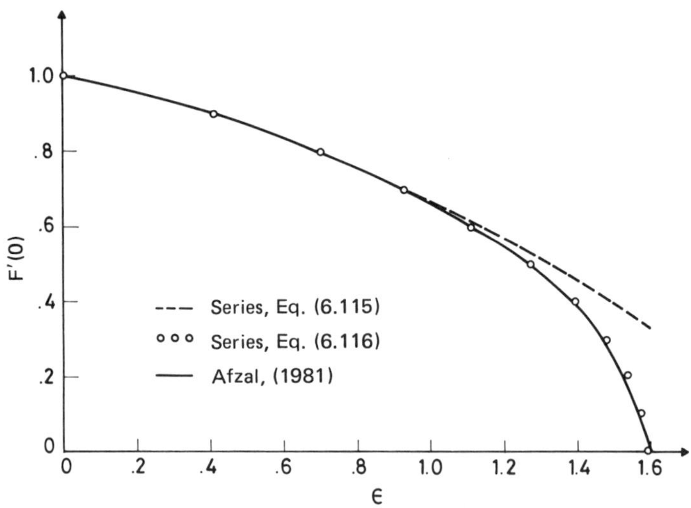

Figure 6.13 Comparison of the series in Eqs. (6.115) and (6.116) and the Afzal (1981) solution.

divergent series. A heuristic motivation for the transformations is discussed by Shanks. Levin (1973) has recently pointed out the shortcomings of the heuristic basis and has suggested further improvements. Out of a half dozen or so transformations that are available, we select the simplest two designated by Shanks as e_1 and e_1^m. The merit of these transformations is that they do not require any information about the analytic structure of the solution. The application is therefore rather blind and this raises the question of how reliable the final results are. This element of unrealiability should not cause undue pessimism because often the pattern of convergence is manifested so convincingly that it speaks for the accuracy of the final results.

Consider first the e_1 transformation. If three partial sums S_{n-1}, S_n, and S_{n+1} of a series are known, then

$$e_1(S_n) = \frac{S_{n+1}S_{n-1} - S_n^2}{S_{n+1} + S_{n-1} - 2S_n} \tag{6.117}$$

The success of e_1 in improving the convergence lies in the fact that, if applied to a geometric series, it yields the exact sum. It is therefore likely to work best on series with nearly geometric coefficients.

The second transformation e_1^m is the mth iteration of e_1. For example, e_1^2 is obtained by treating the sequence $e_1(S_n)$ as partial sums, and so on. Thus, the minimum number of terms needed to apply e_1^2 is five. Similarly, seven terms are needed to continue up to e_1^3, and so on.

We now apply the transformation e_1 and e_1^m to two example series. In the first example, we give the full details but in the second, we quote the main results only. First, consider the series in Eq. (6.85) for $g'(0)$ in laminar natural convection from a nonisothermal vertical plate. As a first step, we calculate the partial sums. These appear in the second column of Table 6.5. The first entry is simply the zero-order term. The second entry is the sum of zero- and first-order terms, namely $-0.358600 - 0.207576 \times 4 = -1.18890$. Similarly, the entry 2.50283 represents the sum of zero-, first-, and second-order terms. The column is continued up to the fifth partial sum which is 53.8724.

Next, we apply e_1 to the entries of the second column. For example, the first entry in the third column is

$$\frac{2.50283 \times (-0.358600) - (-1.18890)^2}{2.50283 - 0.358600 - 2(-1.18890)} = -0.511055$$

Repeating the foregoing calculations with the second set of three numbers, namely -1.18890, 2.50283, and -14.4932, gives the second entry as -0.530114. The process is repeated with the last three numbers to give

Table 6.5 Application of Shanks transformation to the series in Eq. (6.85), $\epsilon = 4$

n	S_n	e_1	e_1^2
0	−0.358600		
1	−1.18890	−0.511055	
2	2.50283	−0.530114	−0.5100
3	−14.4932	−0.881176	
4	53.8724		

−0.881176. Now that we have three numbers in the third column, we can apply the transformation again to give

$$\frac{(-0.511055)\times(-0.88176)-(-0.530114)^2}{-0.511055-0.881176-2(-0.530114)} = -0.5100$$

which appears in the fourth column under e_1^2. Note the table takes the form of a triangular array. The interesting feature is how the wild oscillations of partial sums in the second column are smoothed out with e_1 transformation and refined further with e_1^2 transformation. The final value of −0.5100 compares extremely well with the numerical solution of −0.5065.

We have repeated the calculations for a range of values of ϵ. In Table 6.6 we reproduce a set of results for Pr = 0.73. For comparison, finite difference results (Na, 1978) and asymptotic results (Kuiken, 1969) are also included.

Table 6.6 Comparison of present solution with finite difference and asymptotic solutions for Pr = 0.73

	$-g'(0)$			
ϵ	Present solution	Finite difference (Na, 1978)	Asymptotic solution, small ϵ (Kuiken, 1969)	Asymptotic solution, large ϵ (Kuiken, 1969)
0	0.3586	0.3586	0.3586	
0.25	0.3982	0.3994	0.3916	
0.50	0.4249	0.4262	0.4103	0.4083
0.75	0.4430	0.4433	0.4017	0.4348
1	0.4563	0.4503		0.4517
2	0.4866	0.4848		0.4848
4	0.5100	0.5065		0.5071
6	0.5198	0.5152		0.5157
8	0.5252	0.5199		0.5204
10	0.5286	0.5228		0.5232

Table 6.7 Comparison of Shanks transformed solution with finite difference solution for Pr = 1, 10, and 100

	$-g'(0)$					
	Pr = 1		Pr = 10		Pr = 100	
ϵ	Present solution	Finite difference (Na, 1978)	Present solution	Finite difference (Na, 1978)	Present solution	Finite difference (Na, 1978)
0	0.4011	0.4011	0.8270	0.8270	1.5495	1.5495
0.25	0.4450	0.4457	0.9070	0.9099	1.6955	1.6992
0.50	0.4735	0.4749	0.9583	0.9629	1.7894	1.7930
0.75	0.4932	0.4937	0.9940	0.9990	1.8547	1.8604
1	0.5077	0.5078	1.0203	1.0241	1.9026	1.9035
2	0.5407	0.5391	1.0800	1.0829	2.0112	2.0101
4	0.5663	0.5631	1.1262	1.1295	2.0948	2.0979
6	0.5770	0.5725	1.1456	1.1466	2.1297	2.1263
8	0.5829	0.5777	1.1563	1.1567	2.1488	2.1447
10	0.5867	0.5809	1.1630	1.1630	2.1609	2.1563

The twice Shanks transformed solution agrees to within one percent up to $\epsilon = 10$, which effectively represents the asymptotic limit of $\epsilon \to \infty$ (Kuiken, 1969). The same remarkable accuracy is evidenced at other Prandtl numbers as shown in Table 6.7.

For the second example, consider the series in Eq. (6.55). A sample application of iterated Shanks transformation e_1^m is shown in Table 6.8. Since we have eleven terms we are able to iterate up to ϵ^5. As in the previous example,

Table 6.8 Iterated application of Shanks transformation to the series in Eq. (6.55), $\epsilon = 1$

n	S_n	e_1	e_1^2	e_1^3	e_1^4	e_1^5
0	0.469600					
1	1.768529	1.067675				
2	0.246380	1.312898	1.210872			
3	3.809349	1.138180	1.241830	1.230621		
4	−6.862651	1.392997	1.224260	1.233572	1.232539	
5	29.599035	0.893491	1.244073	1.231982	1.232697	1.232623
6	−105.345958	2.069371	1.213054	1.233280	1.232560	
7	421.182851	−1.081582	1.271184	1.231184		
8	−1711.227061	8.206277	1.147714			
9	7167.089345	−21.201637				
10	−30595.609874					

the wildly oscillating partial sums of the second column are progressively smoothed out as the transformation is iterated. The last column gives a value of $f''(0) = 1.232623$ which is in excellent agreement with the value of 1.232588 of Smith (1954) and 1.232561 of Cebeci and Keller (1971). Indeed, this level of accuracy is achieved for the whole range of $-0.19884 \leqslant \epsilon \leqslant 2$ as shown in Table 6.9.

The series in Eq. (6.56) for $-g'(0)$ can be similarly improved. The final results are collected in Table 6.10 where a close agreement between the present results and exact solution is exhibited.

6.4.5 Padé Approximants

The technique of forming Padé approximants to accelerate convergence of a power series has been extensively used in physics and chemistry, particularly in the field of cooperative phenomenon and critical points. However, its application in engineering has been limited (Guttman, 1975; Aziz, 1981). The survey article of Baker (1965) gives an excellent exposition of the subject.

The Padé approximant of a series in the form of the ratio of two polynomials. For example, the Padé approximant $[N, M]$ is the fraction P/Q

Table 6.9 Comparison of values of $f''(0)$

	$f''(0)$		
ϵ	Smith (1954)	Cebeci & Keller (1971)	Aziz & Na (1981)
2.0	1.687218	–	1.687516
1.6	1.521514	1.521516	1.521689
1.2	1.335722	1.335724	1.335793
1.0	1.232588	1.232561	1.232623
0.8	1.120268	1.120269	1.120280
0.6	0.995836	–	0.995837
0.4	0.854421	0.854423	0.854418
0.2	0.686708	0.686711	0.686706
0.1	0.587035	0.587037	0.587034
0.05	0.531130	–	0.531129
0.00	0.469600	0.469603	0.469600
–0.05	0.400323	0.400330	0.400322
–0.10	0.319270	0.319278	0.319266
–0.14	0.239736	–	0.239724
–0.16	0.190780	–	0.190758
–0.18	0.128636	–	0.128615
–0.19	0.085700	0.085702	0.085840
–0.195	0.055172	0.055177	0.056027

Table 6.10 Comparison of values of $-g'(0)$, Pr = 0.72

ϵ	Present	Exact
2.0	0.529361	0.529599
1.6	0.520296	0.520441
1.2	0.508646	0.508722
0.8	0.492697	0.492726
0.6	0.482006	0.482019
0.4	0.468221	0.468223
0.2	0.449063	0.449057
−0.05	0.406238	0.406223
−0.10	0.390730	0.390714
−0.16	0.361395	0.361330
−0.18	0.344201	0.344158
−0.19	0.330886	0.330686
−0.195	0.320839	0.320071

where P is a polynomial of degree M and Q is a polynomial of degree N. The coefficients of these polynomials are obtained by imposing the condition that the expansion of the Padé approximant and the original series must agree to the order $M + N$, and $Q(0) = 1$.

As remarked by Van Dyke (1974), the diagonal approximants where $M = N$ are most useful in improving the convergence. However, like the Shanks transformation, the approximants are formed without any knowledge of the analytic structure, and therefore their accuracy is not always reliable. Nonetheless, the convergence can often be confirmed by forming the sequence [1, 1], [2, 2], [3, 3], and so on.

Since the zeros of polynomial Q simulate the singularities of the function, the Padé approximants prove most useful when the coefficients exhibit irregular pattern or the radius of convergence is zero.

For the purpose of illustration, let us derive the [1, 1] Padé approximant for the series $\Sigma_{n=0}^{\infty} \epsilon^n a_n$. Let the first-degree polynomial for the numerator be $P = b_0 + b_1 \epsilon$, and that for the denominator be $Q = c_0 + c_1 \epsilon$ where $c_0 = 1$ to meet the condition $Q(0) = 1$. The fraction P/Q must agree to (1 + 1), i.e., second-degree with the original series. Thus

$$[1, 1] = \frac{b_0 + b_1 \epsilon}{1 + c_1 \epsilon} = a_0 + a_1 \epsilon + a_2 \epsilon^2 \tag{6.118}$$

To obtain b_0, b_1, and c_1, we write the above as

$$(b_0 + b_1 \epsilon)(1 + c_1 \epsilon)^{-1} = a_0 + a_1 \epsilon + a_2 \epsilon^2$$

and then expand $(1 + c_1 \epsilon)^{-1}$ binomially to give

$$(b_0 + b_1 \epsilon)(1 - c_1 \epsilon + c_1^2 \epsilon^2 + \cdots) = a_0 + a_1 \epsilon + a_2 \epsilon^2$$

Expanding and equating coefficients of ϵ^0, ϵ^1, and ϵ^2 on both sides, we get

$$\begin{aligned} b_0 &= a_0 \\ b_1 - b_0 c_1 &= a_1 \\ b_0 c_1^2 - b_1 c_1 &= a_2 \end{aligned} \tag{6.119}$$

Solving Eq. (6.119) for b_0, b_1, and c_1, we have

$$b_0 = a_0$$

$$b_1 = \frac{a_1^2 - a_0 a_2}{a_1}$$

$$c_1 = -\frac{a_2}{a_1}$$

Thus from Eq. (6.118) it follows that

$$[1, 1] = \frac{a_0 + [(a_1^2 - a_0 a_2)/a_1]\epsilon}{1 - (a_2/a_1)\epsilon} = \frac{a_0 a_1 + (a_1^2 - a_0 a_2)\epsilon}{a_1 - a_2 \epsilon} \tag{6.120}$$

It should be noted that to evaluate the coefficient of approximants $[N, N]$, we need to know $(2N + 1)$ coefficients of the original series. Thus with just three terms, one can only form the first diagonal approximant $[1, 1]$. With five terms, one can form the second diagonal approximant $[2, 2]$, which is given by

$$[2, 2] = \frac{P}{Q} \tag{6.121}$$

where

$$\begin{aligned} P = \{&a_0(a_2^2 - a_1 a_3) + [a_1(a_2^2 - a_1 a_3) + a_0(a_1 a_4 - a_2 a_3)]\epsilon \\ &+ [a_0(a_3^2 - a_2 a_4) + a_1(a_1 a_4 - a_2 a_3) + a_2(a_2^2 - a_1 a_3)]\epsilon^2\} \end{aligned}$$

$$Q = a_2^2 - a_1 a_3 + (a_1 a_4 - a_2 a_3)\epsilon + (a_3^2 - a_2 a_4)\epsilon^2$$

Indeed one can generate a whole set of Padé approximants and arrange them in a Padé table as shown below.

[0, 0]	[0, 1]	[0, 2]	⋯
[1, 0]	[1, 1]	[1, 2]	⋯
[2, 0]	[2, 1]	[2, 2]	⋯
⋯	⋯	⋯	⋯

Without entering into further discussion which can be found in Baker (1965), we will illustrate with two examples the usefulness of $[2, 2]$ approximants in accelerating convergence.

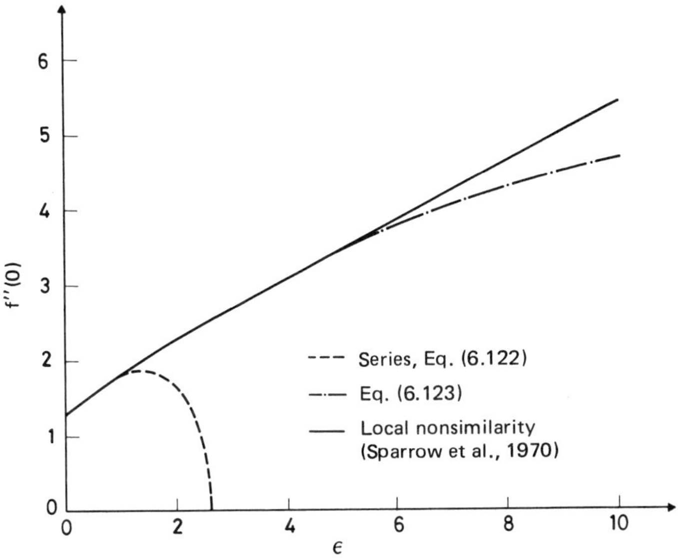

Figure 6.14 Comparison of the series in Eqs. (6.40) and (6.123) and the local nonsimilarity solution.

First, consider the series in Eq. (6.40). Retaining the first five terms, we have

$$f''(0) = 1.3282 + 0.6943\epsilon - 0.1641\epsilon^2 + 0.1016\epsilon^3 - 0.07595\epsilon^4 \tag{6.122}$$

Forming [2, 2] approximant from Eq. (6.122) we get

$$f''(0) = \frac{0.0579 + 0.0782\epsilon + 0.0207\epsilon^2}{0.0436 + 0.0361\epsilon + 0.0021\epsilon^2} \tag{6.123}$$

In Fig. 6.14 we plot Eqs. (6.122) and (6.123) for a range of ϵ from 0 to 10 and compare with the local nonsimilarity solution of Sparrow et al. (1970). Whereas the original series in Eq. (6.122) diverges beyond $\epsilon = 1$, the [2, 2] approximant given by Eq. (6.123) is accurate up to $\epsilon = 5$, which means the range is extended at least five times. Interestingly enough, the application of iterated Shanks transformation to nine terms of the series in Eq. (6.40) gives a solution which when compared with the local nonsimilarity solution is remarkably accurate right up to $\epsilon = 40$ (see Problem 6.11).

Second, consider the series in Eq. (6.17). Forming [2, 2] approximant from the first five terms, we have

$$\theta(0) = \frac{0.258334 + 0.804399\epsilon + 0.319895\epsilon^2}{0.258334 + 0.933566\epsilon + 0.571401\epsilon^2} \tag{6.124}$$

The results for fin efficiency, calculated using Eqs. (6.124) and (6.106), are almost identical to those obtained using the Eulerized series in Eqs.

(6.105) and (6.106). Both results agree remarkably well with the direct numerical solution in the range $\epsilon = 0$ to 4, which covers the entire range of fin operation in practice.

PROBLEMS

6.1 For transient conduction in a semi-infinite medium with exponential type thermal diffusivity dependence on temperature, the one-dimensional heat equation becomes

$$\theta''(1-\epsilon\theta) + 2\eta\theta' = 0$$

$$\theta(0) = 1 \qquad \theta'(\infty) = 0$$

Assuming a perturbation expansion, derive the first eleven perturbation equations. Solve them numerically and show that

$$\begin{aligned}-\theta'(0) = {} & 1.12837917 + 0.20501534\epsilon + 0.08897813\epsilon^2 + 0.05054114\epsilon^3 \\ & + 0.03287074\epsilon^4 + 0.02321012\epsilon^5 + 0.01732255\epsilon^6 \\ & + 0.01345554\epsilon^7 + 0.01077242\epsilon^8 + 0.00883088\epsilon^9 \\ & + 0.00737851\epsilon^{10}\end{aligned}$$

6.2 Consider Problem 2.12. Show that the sequence of perturbation equations is given by

$$\sum_{n=0}^{\infty} \epsilon^n(\theta_n'' + 2\eta\theta_n' - 4n\theta_n) = \sum_{n=0}^{\infty} \epsilon^{n+1}\theta_n$$

Extend analytically the three-term solution given in Problem 2.12 to five terms showing that θ_3 and θ_4 have the solutions

$$\theta_3 = i^6 \operatorname{erfc}\eta - \frac{1}{4} i^4 \operatorname{erfc}\eta + \frac{1}{32} i^2 \operatorname{erfc}\eta - \frac{1}{384}\operatorname{erfc}\eta$$

$$\theta_4 = i^8 \operatorname{erfc}\eta - \frac{1}{4} i^6 \operatorname{erfc}\eta + \frac{1}{32} i^4 \operatorname{erfc}\eta - \frac{1}{384} i^2 \operatorname{erfc}\eta + \frac{1}{6144}\operatorname{erfc}\eta$$

6.3 Consider Problem 5.7. Carry out a perturbation expansion in inverse powers of ϵ, and extend the solution for θ to five terms. The solution for the first three terms is given by Malmuth et al. (1970). Derive the five-term solution for $\theta(0)$ as (Aziz, 1981)

$$\begin{aligned}\theta(0) = {} & 0.90660600 - 0.0092962\epsilon^{-1} + 0.00078384\epsilon^{-2} \\ & - 0.00008506\epsilon^{-3} + 0.0001021\epsilon^{-4}\end{aligned}$$

6.4 Na and Chiou (1979) consider the problem of natural convection over a frustum of a cone. For constant surface temperature, the equations are

$$f''' + \left(\frac{\epsilon}{1+\epsilon} + \frac{3}{4}\right) ff'' - \frac{1}{2}(f')^2 + g = \epsilon \left(f' \frac{\partial f'}{\partial \epsilon} - f'' \frac{\partial f}{\partial \epsilon}\right)$$

$$\frac{1}{\Pr} g'' + \left(\frac{\epsilon}{1+\epsilon} + \frac{3}{4}\right) fg' = \epsilon \left(f' \frac{\partial g}{\partial \epsilon} - g' \frac{\partial f}{\partial \epsilon}\right)$$

$$f(\epsilon, 0) = f'(\epsilon, 0) = 0 \qquad f'(\epsilon, \infty) = 0$$

$$g(\epsilon, 0) = 1 \qquad g(\epsilon, \infty) = 0$$

Derive the first five perturbation equations for f and g. Solve them for $\Pr = 10$ and verify that

$$-g'(0) = 0.827000 + 0.155652\epsilon - 0.114565\epsilon^2 + 0.101101\epsilon^3 - 0.077713\epsilon^4$$

6.5 Considering the problem of heat transfer through a draining film, Isenberg and Gutfinger (1973) transformed the energy equation into

$$\theta'' + \psi\theta' = \frac{1}{2}\epsilon(3 + \epsilon\psi^2)\frac{\partial \theta}{\partial \epsilon}$$

$$\psi = 0 \qquad \theta = 0$$

$$\psi = \infty \qquad \theta = 1$$

where primes denote differentiation with respect to ψ. Expand θ in a perturbation series in ϵ and derive the governing equations for θ_0 to θ_4. Solve the sequence of equations to obtain $\theta'(0)$ series up to $O(\epsilon^4)$.

6.6 The combined forced and free convection from a vertical thin needle in a variable external stream is considered by Narain and Uberoi (1973). The boundary-layer equations are reduced to

$$8\eta f''' + 8f'' + 4ff'' - 2f'^2 + \epsilon\theta + \tfrac{1}{2} = 0$$

$$\eta\theta'' + (1 + \tfrac{1}{2}\Pr f)\theta' = 0$$

$$f(a) = f'(a) = 0 \qquad f(\infty) = \tfrac{1}{2}$$

$$\theta(a) = 1 \qquad \theta(\infty) = 0$$

where primes denote differentiation with respect to η. Expand f and θ in a power series of ϵ and solve for the first eleven terms of the series.

6.7 In an analysis of the effect of uniform suction on laminar film condensation on a porous vertical wall, Yang (1969) writes the nonsimilar boundary-layer equations as

$$f''' + 3ff'' - 2(f')^2 + 1 = \epsilon \left(f'' + f' \frac{\partial f'}{\partial \epsilon} - f'' \frac{\partial f}{\partial \epsilon} \right)$$

$$\theta'' + 3\,\mathrm{Pr}\, f\theta' + \epsilon\, \mathrm{Pr} \left(\theta' \frac{\partial f}{\partial \epsilon} - f' \frac{\partial \theta}{\partial \epsilon} - \theta' \right) = 0$$

$$\eta = 0 \quad f' = 0 \quad 3f + \epsilon \frac{\partial f}{\partial \epsilon} = 0 \quad \theta = 1$$

$$\eta = \eta_\delta \quad f'' = 0 \quad \theta = 0$$

where primes denote differentiation with respect to η. Assuming a perturbation series for f and θ, deduce the sequence of perturbation equations up to $O(\epsilon^{10})$,

6.8 The cooling of a low heat resistance stretching sheet moving through a fluid is described by the following model (Afzal and Varshney, 1980)

$$f''' + ff'' - \beta f'^2 = 0$$

$$\frac{1}{\mathrm{Pr}} \theta'' + f\theta' = \epsilon(1 - \beta) f' \frac{\partial \theta}{\partial \epsilon}$$

$$f(0) = 0 \quad f'(0) = 1 \quad f'(\infty) = 0$$

$$\theta(0, \epsilon) = 1 - \epsilon \int_0^\infty f\theta \, d\eta$$

$$\theta(\infty, \epsilon) = 0$$

For small ϵ, expand f and θ into perturbation series, and evaluate numerically for $\beta = 0.5$ the first eleven terms of the series for $\theta(0, \epsilon)$. Compare your solution with the Afzal and Varshney (1980) solution which reads

$$\begin{aligned}\theta(0, \epsilon) = 1 &- 0.6682\epsilon + 0.41410\epsilon^2 - 0.23999\epsilon^3 + 0.13108\epsilon^4 \\ &- 0.06787\epsilon^5 + 0.03349\epsilon^6 - 0.01580\epsilon^7 + 0.00716\epsilon^8 \\ &- 0.00312\epsilon^9 + 0.00132\epsilon^{10}\end{aligned}$$

6.9 In calculating heat transfer through compressible laminar boundary-layer, Lighthill (1950) derived the following integral equation

$$[F(\epsilon)]^4 = -\frac{1}{2} \epsilon^{-1/2} \int_0^\epsilon \frac{F'(u)\, du}{(\epsilon^{3/2} - u^{3/2})^{1/3}}$$

$$F(0) = 1$$

Assuming a perturbation expansion for $F(\epsilon)$ show that (Curle, 1978)

$$F(\epsilon) = 1 - 1.460998\epsilon + 7.2494161\epsilon^2 - 46.449738\epsilon^3 + 332.75523\epsilon^4 - 2536.8206\epsilon^5 + 20120.061\epsilon^6 - 16399.85\epsilon^7 + 1363564.3\epsilon^8 - 11511356\epsilon^9$$

6.10 Derive the eleven-term series for the positive root of $x^2 + x - \epsilon = 0$. Draw a Domb-Sykes plot and estimate ϵ_0. Next, apply Euler transformation to obtain the improved series. Assess the accuracy of the original and improved series in light of the exact solution.

6.11 Use the series in Eq. (6.40) for the skin friction in longitudinal flow over a cylinder and apply iterated Shanks transformation up to e_1^4. Carry out the computations for a range of values of ϵ from 0 to 40. Verify the information in the following table

ϵ	$f''(\epsilon, 0)$
1	1.91753
4	3.06060
10	5.43390
20	8.46835
30	11.07511
40	13.35569

Converting to the variables used by Sparrow et al. (1970), that is, $x^* = 2\log_{10}(\epsilon/4)$ and $y^* = \log_{10}[2\pi f''(\epsilon, 0)/\epsilon]$, demonstrate that the iterated Shanks transformation results and those obtained by local nonsimilarity method are in excellent agreement.

6.12 Apply the Shanks transformation e_1 to the series in Eqs. (4.110) and (4.116) to show that

$$e_1(r_f) = \psi\,\frac{60\psi^2 + \epsilon(32\psi - 13)}{60\psi^2 + \epsilon(22\psi - 3)}$$

$$e_1(\tau) = \frac{[60\psi^2(1 + 2\psi) + \epsilon(1 + 2\psi + 120\psi^2)](1 - \psi)^2}{6(60\psi^2 + \epsilon)}$$

6.13 Consider the result quoted in Problem 6.3. Take $\epsilon = 0.01$ and show that the twice Shanks transformed result gives $\theta(0) = 0.7723$. This result is only 0.5 percent in error compared to the numerical solution of 0.7681 (Aziz, 1981).

6.14 Draw a Domb-Sykes plot for the series for $\theta(0, \epsilon)$ given in Problem 6.8. Estimate the intercept ϵ_0 and slope α. Try different improvement schemes and compare the results with those given by Afzal and Varshney (1980).

6.15 Refer to the series for $-g'(0)$ in Problem 6.4. Apply the Shanks transformation twice to produce the results tabulated below. For comparison, the finite-difference solutions have also been tabulated.

	$-g'(0)$, Pr = 10	
ϵ	Twice Shanks transformed solution	Finite-difference solution
0	0.8270	0.8269
0.25	0.8601	0.8634
0.75	0.9078	0.9029
3.75	0.9456	0.9823
7.75	0.9709	1.0087
15.75	0.9844	1.0184
31.75	0.9916	1.0285
63.75	0.9953	1.0289
127.75	0.9972	1.0337

6.16 Using Eq. (6.89) and Table 6.2, write the eleven-term series for $F'(0)$ for the case of assisting flow. Draw a Domb-Sykes plot to estimate ϵ_0 (cf. $\epsilon_0 = 1.6$). Hence obtain the Eulerized series. Compare your results with those reported by Afzal (1981).

6.17 Obtain a [2, 2] Padé approximant for the series for $\theta(0)$ appearing in Problem 6.3. Using a value of $\epsilon = 0.01$, calculate $\theta(0)$ and compare the answer with the values quoted in Problem 6.13.

6.18 For the series given in Problem 6.4, form a [2, 2] Padé approximant and compare its predictions with the finite-difference and the twice Shanks transformed solutions given in Problem 6.15. Discuss the efficacy of [2, 2] approximant versus the Shanks transformation.

REFERENCES

Afzal, N. (1981). Mixed Convection in a Two-dimensional Buoyant Plume, *J. Fluid Mech.*, vol. 105, pp. 347–368.

Afzal, N. and Varshney, I. S. (1980). The Cooling of a Low Heat Resistance Stretching Sheet Moving Through a Fluid, *Wärme Stoffübertrag.*, vol. 14, pp. 289–293.

Anderson, A. D. and Reynolds, W. C. (1970). Perturbation Analysis of the Quasi-Developed Flow of a Liquid in a Horizontal Tube, Tech. Rept. FM-7, Stanford University, California.

Arunachalam, M. and Rajappa, N. R. (1978). Thermal Boundary Layer in Liquid Metals with Variable Thermal Conductivity, *Appl. Sci. Res.*, vol. 34, pp. 179–187.

Arunachalam, M. and Seeniraj, V. (1977). Comments on the paper "Application of Perturbation Techniques to Heat Transfer Problems with Variable Thermal Properties," *Int. J. Heat Mass Transfer*, vol. 20, pp. 803–804.

Asfar, O. R., Aziz, A., and Soliman, M. A. (1979). A Uniformly Valid Solution for Inward Cylindrical Solidification, *Mech. Res. Comm.*, vol. 6, pp. 325–332.

Aziz, A. (1977). Perturbation Solutions of Certain Transcendental Equations in Heat Conduction, *Mech. Eng. News (ASEE)*, vol. 14, pp. 2–5.

Aziz, A. (1978). New Asymptotic Solution for the Variable Property Stefan Problem, *Chem. Eng. J.*, vol. 16, pp. 65–68.

Aziz, A. (1979). Radiating Fin Analysis with Extended Perturbation Series, *Lett. Heat Mass Transfer*, vol. 6, pp. 199–203.

Aziz, A. (1981). Analysis of Heat Transfer Problems with Computer Extended Perturbation Series, *Num. Heat Transfer*, vol. 4, pp. 123–130.

Aziz, A. and Benzies, J. Y. (1976). Application of Perturbation Techniques to Heat Transfer Problems with Variable Thermal Properties, *Int. J. Heat Mass Transfer*, vol. 19, pp. 271–276.

Aziz, A. and El-Ariny, A. S. (1977). A Perturbation Approach to Heat Conduction with Temperature Dependent Generation and Thermal Conductivity, ASME Paper No. 77-HT-28.

Aziz, A. and Huq, E. (1975). Perturbation Solution for Convecting Fin with Variable Thermal Conductivity, *J. Heat Transfer*, vol. 97, pp. 300–301.

Aziz, A. and Hamad, G. (1977). Regular Perturbation Expansions in Heat Transfer, *Int. J. Mech. Eng. Educ.*, vol. 5, pp. 167–182.

Aziz, A., Jaleel, A., and Haneef, M. (1980). Perturbation Methods in Heat Conduction, *Chem. Eng. J.*, vol. 19, pp. 171–182.

Aziz, A. and Na, T. Y. (1980). Transient Response of Fins by Coordinate Perturbation Expansions, *Int. J. Heat Mass Transfer*, vol. 23, pp. 1695–1698.

Aziz, A. and Na, T. Y. (1981). New Approach to the Solution of Falkner-Skan Equation, *AIAA J.*, vol. 19, pp. 1242–1244.

Baker, G. A., Jr. (1965). The Theory and Application of the Padé Approximant Method, in K. A. Brueckner (ed.), *Advances in Theoretical Physics*, Academic, New York, pp. 1–58.

Beckett, P. M. (1981). On the Use of Series Solution Applied to Solidification Problems, *Mech. Res. Comm.*, vol. 8, pp. 169–174.

Bellman, R. (1964). *Perturbation Techniques in Mathematics, Physics and Engineering*, Holt, Rinehart & Winston, New York.

Carslaw, H. S. and Jaeger, J. C. (1959). *Conduction of Heat in Solids*, The Clarendon Press, Oxford.

Cebeci, T. and Bard, J. (1973). Thermal Response of an Unsteady Laminar Boundary Layer on a Flat Plate due to Step Changes in Wall Temperature and in Wall Heat Flux, *Comput. Meth. Appl. Mech. Eng.*, vol. 2, pp. 323–338.

Cebeci, T. and Keller, H. B. (1971). Shooting and Parallel Shooting Methods for Solving the Falkner-Skan Boundary Layer Equation, *J. Comput. Phys.*, vol. 7, pp. 289–300.

Cess, R. D. (1966). Interaction of Thermal Radiation with Free Convection Heat Transfer, *Int. J. Heat Mass Transfer*, vol. 9, pp. 1269–1277.

Chen, C. J., Naseri-Neshat, H., and Li, P. (1980). The Finite Analytic Method, Report E-CJC-1-80, Energy Division and Iowa Institute of Hydraulic Research, The University of Iowa, Iowa City.

Cohen, D. S. and Shair, F. H. (1970). Steady State Temperature Profiles Within Insulated Electrical Cables Having Variable Conductivities, *Int. J. Heat Mass Transfer*, vol. 13, pp. 1375–1378.

Cole, J. D. (1968). *Perturbation Methods in Applied Mathematics*, Blaisedell, Waltham, MA.

Curle, N. (1978). Solution of an Integral Equation of Lighthill, *Proc. R. Soc. London*, A. 364, pp. 435–441.

Domb, C. and Sykes, M. F. (1957). On the Susceptibility of a Ferromagnetic Above the Curie Point, *Proc. R. Soc. London*, A. 240, pp. 214–228.

Erdélyi, A. (1956). *Asymptotic Expansions*, Dover, New York.

Gebhart, B. (1962). Effect of Viscous Dissipation in Natural Convection, *J. Fluid Mech.*, vol. 14, pp. 225–232.

Goodman, T. R. and Shea, J. J. (1960). The Melting of Finite Slabs, *J. Appl. Mech.*, vol. 82, pp. 16–24.

Gorla, R. S. R. (1976). Heat Transfer in an Axisymmetric Stagnation Flow on a Cylinder, *Appl. Sci. Res.*, vol. 32, pp. 541–553.

Gray, D. D. (1977). The Laminar Plume Above a Line Heat Source in a Transverse Magnetic Field, *Appl. Sci. Res.*, vol. 33, pp. 437–453.

Guttman, A. J. (1975). Derivation of "Mimic Functions" from Regular Perturbation Expansions in Fluid Mechanics, *J. Inst. Math. Its Appl.*, vol. 15, pp. 307–317.

Hering, R. G. (1965). Laminar Free Convection from a Non-Isothermal Cone at Low Prandtl Numbers, *Int. J. Heat Mass Transfer*, vol. 8, pp. 1333–1337.

Huang, C. L. and Shih, Y. P. (1975). Perturbation Solution for Planar Solidification of a Saturated Liquid with Convection at the Wall, *Int. J. Heat Mass Transfer*, vol. 18, pp. 1481–1483.

Imber, M. (1979). Nonlinear Heat Transfer in Planar Solids: Direct and Inverse Applications, *AIAA J.*, vol. 17, pp. 204–212.

Ingham, D. B. (1977). Singular Parabolic Partial Differential Equations that Arise in Impulsive Motion Problems, *J. Appl. Mech.*, vol. 99, pp. 396–400.

Isenberg, J. and Gutfinger, C. (1973). Heat Transfer to a Draining Film, *Int. J. Heat Mass Transfer*, vol. 16, pp. 505–512.

Krane, R. J., Jischke, M. C., and Rasmussen, M. L. (1973). The Thermal Analysis of a Belt Type Radiator by the Method of Matched Asymptotic Expansions, *Int. J. Heat Mass Transfer*, vol. 16, pp. 1165–1174.

Kuiken, H. K. (1968a). An Asymptotic Solution for Large Prandtl Number Free Convection, *J. Eng. Math.*, vol. 2, pp. 355–371.

Kuiken, H. K. (1968b). Axisymmetric Free Convection Boundary Layer Flow Past Slender Bodies, *Int. J. Heat Mass Transfer*, vol. 11, pp. 1141–1153.

Kuiken, H. K. (1969). General Series Solution for Free Convection Past a Nonisothermal Vertical Flat Plate, *Appl. Sci. Res.*, vol. 20, pp. 205–215.

Kuo, Y. H. (1953). On the Flow of an Incompressible Viscous Fluid Past a Flat Plate at Moderate Reynolds Numbers, *J. Math. Phys.*, vol. 32, pp. 83–101.

Kuo, Y. H. (1956). Viscous Flow Along a Flat Plate Moving at High Supersonic Speeds, *J. Aeron. Sci.*, vol. 23, pp. 125–136.

Levin, D. (1973). Development of Nonlinear Transformations for Improving Convergence of Sequences, *Int. J. Comput. Math.*, vol. 3, pp. 371–388.

Lighthill, M. J. (1949). A Technique for Rendering Approximate Solutions to Physical Problems Uniformly Valid, *Philos. Mag.*, vol. 40, pp. 1179–1201.

Lighthill, M. J. (1950). Contributions to the Theory of Heat Transfer Through a Laminar Boundary Layer, *Proc. R. Soc. Lond.*, A. 202, pp. 359–386.

Linstedt, A. (1882). Ueber Die Integration Einer Fur Die Strorungstheorie Wichtigen Differentialgleichung, *Astron. Nach.*, vol. 103, col. 211–220.

Malmuth, M., Kassic, M., and Mueller, H. F. (1970). Asymptotic and Numerical Solutions for Nonlinear Conduction in Radiating Heat Shields, *J. Heat Transfer*, vol. 92, pp. 264–268.

Martin, E. D. (1967). Simplified Application of Lighthill's Uniformization Technique Using Lagrange Expansion Formulas, *J. Inst. Math. Appl.*, vol. 3, pp. 16–20.

Minkowycz, W. J. and Sparrow, E. M. (1974). Local Nonsimilar Solutions for Natural Convection on a Vertical Cylinder, *J. Heat Transfer*, vol. 96, pp. 178–183.

Morton, B. R. (1959). Laminar Convection in Uniformly Heated Horizontal Pipes at Low Rayleigh Numbers, *Q. J. Mech. Appl. Math.*, vol. 12, pp. 410–420.

Morton, B. R. (1960). Laminar Convection in Uniformly Heated Vertical Pipes, *J. Fluid Mech.*, vol. 8, pp. 227–240.

Mueller, H. F. and Malmuth, N. D. (1965). Temperature Distribution in Radiating Heat Shields by the Method of Singular Perturbations, *Int. J. Heat Mass Transfer*, vol. 8, pp. 915–920.

Na, T. Y. (1978). Numerical Solution of Natural Convection Flow Past a Nonisothermal Vertical Flat Plate, *Appl. Sci. Res.*, vol. 33, pp. 519–543.

Na, T. Y. (1979). *Computational Methods in Engineering Boundary Value Problems*, Academic, New York.

Na, T. Y. and Chiou, J. P. (1979). Laminar Natural Convection Over a Frustum of a Cone, *Appl. Sci. Res.*, vol. 35, pp. 409–421.

Narain, J. P. and Uberoi, M. S. (1973). Combined Forced and Free Convection Over Thin Needles, *Int. J. Heat Mass Transfer*, vol. 16, pp. 1505–1512.

Nayfeh, A. H. (1973). *Perturbation Methods*, Wiley, New York.

Nayfeh, A. H. (1981). *Introduction to Perturbation Techniques*, Wiley, New York.

O'Malley, R. E., Jr. (1974). *Introduction to Singular Perturbations*, Academic, New York.

Pedroso, R. I. and Domoto, G. A. (1973a). Perturbation Solutions for Spherical Solidification of Saturated Liquids, *J. Heat Transfer*, vol. 95, pp. 42–46.

Pedroso, R. I. and Domoto, G. A. (1973b). Exact Solution by Perturbation Method for Planar Solidification of a Saturated Liquid with Convection at the Wall, *Int. J. Heat Mass Transfer*, vol. 16, pp. 1816–1819.

Pedroso, R. I. and Domoto, G. A. (1973c). Inward Spherical Solidification–Solution by the Method of Strained Coordinates, *Int. J. Heat Mass Transfer*, vol. 16, pp. 1037–1043.

Pedroso, R. I. and Domoto, G. A. (1973d). Technical Note on Planar Solidification with Fixed Wall Temperature and Variable Thermal Properties, *J. Heat Transfer*, vol. 93, pp. 553–555.

Poincaré, H. (1982). *New Methods of Celestial Mechanics*, Vol. I–III (English transl.), NASA TTF-450, 1967.

Pritulo, M. F. (1962). On the Determination of Uniformly Accurate Solutions of Differential Equations by the Method of Perturbation of Coordinates, *J. Appl. Math. Mech.*, vol. 26, pp. 661–667.

Riley, N. (1963). Unsteady Heat Transfer for Flow Over a Flat Plate, *J. Fluid Mech.*, vol. 17, pp. 97–104.

Roy, S. (1969). A Note on Natural Convection at High Prandtl Numbers, *Int. J. Heat Mass Transfer*, vol. 12, pp. 239–241.

Scheffler, W. A. (1974). Heat Conduction in Bodies with Small Boundary Perturbations, *J. Heat Transfer*, vol. 96, pp. 248–250.

Schenk, J., Altmann, R., and Dewit, J. P. A. (1976). Interaction between Heat and Mass Transfer in Simultaneous Natural Convection About an Isothermal Vertical Flat Plate, *Appl. Sci. Res.*, vol. 32, pp. 599–605.

Seban, R. A. and Bond, R. (1951). Skin Friction and Heat Transfer Characteristics of a Laminar Boundary Layer on a Cylinder in Axial Incompressible Flow, *J. Aeronaut. Sci.*, vol. 18, pp. 671–675.

Shanks, D. (1955). Nonlinear Transformations of Divergent and Slowly Convergent Sequences, *J. Math. Phys.*, vol. 34, pp. 1–42.

Shih, Tsan-Hsing (1981). Computer-Extended Series: Natural Convection in a Long Horizontal Pipe with Different End Temperatures, *Int. J. Heat Mass Transfer*, vol. 24, pp. 1295–1303.

Shih, Y. P. and Ju, Shin-Jon (1976). Perturbation Solution of Inverse Stefan Problem for Forced Flow Inside or Outside Cylinders, *J. Chin. Inst. Chem. Eng.*, vol. 7, pp. 53–58.

Shih, Y. P. and Tsou, J. D. (1978). Extended Leveque Solutions for Heat Transfer to Power Law Fluids in Laminar Flow in a Pipe, *Chem. Eng. J.*, vol. 15, pp. 55–62.

Singh, A. K. (1979). Boundary Layer Flows with Swirl and Large Suction, *Appl. Sci. Res.*, vol. 35, pp. 59–65.

Smith, A. M. O. (1954). Improved Solutions of Falkner-Skan Boundary Layer Equation, Fund Paper, *J. Aero. Sci.*

Solomon, A. D. (1979). Mathematical Modelling of Phase Change Processes for Latent Heat Thermal Energy Storage, Union Carbide Corporation, Report No. ORNL/CSD-39.

Sparrow, E. M. and Cess, R. D. (1978). *Radiation Heat Transfer*, Augmented Edition, Hemisphere, Washington, D.C.

Sparrow, E. M. and Niewerth, E. R. (1968). Radiating, Convecting and Convecting Fins: Numerical and Linearized Solutions, *Int. J. Heat Mass Transfer*, vol. 11, pp. 377-379.

Sparrow, E. M., Quack, H., and Boerner, C. J. (1970). Local Nonsimilarity Boundary-Layer Solutions, *AIAA J.*, vol. 8, pp. 1936–1942.

Stephan, K. and Holzknecht, B. (1974). Heat Conduction in Solidification of Geometrically Simple Bodies, *Wärme Stoffübertrag.*, vol. 7, pp. 200–207.

Tao, L. C. (1967). Tabulation of Numerical Solutions of Freezing a Saturated Liquid in a Cylinder or a Sphere, *Document 9159*, American Documentation Institute, Library of Congress, Washington, D.C.

Tsien, H. S. (1956). The Poincaré-Lighthill-Kuo Method, *Adv. Appl. Mech.*, vol. 4, pp. 281–349.

Turian, R. M. and Bird, R. B. (1963). Viscous Heating in the Cone-and-Plate Viscometer-II, *Chem. Eng. Sci.*, vol. 18, pp. 689–696.

Van Dyke, M. (1974). Analysis and Improvement of Perturbation Series, *Q. J. Mech. Appl. Math.*, vol. 27, pp. 423–450.

Van Dyke, M. (1975a). *Perturbation Methods in Fluid Mechanics*, Annotated Edition, Parabolic Press, Stanford, California.

Van Dyke, M. (1975b). Computer Extension of Perturbation Series in Fluid Mechanics, *SIAM J. Appl. Math.*, vol. 28, pp. 720–734.

Van Dyke, M. (1977). From Zero to Infinite Reynolds Number by Computer Extension of Stokes Series, in A. Dold and B. Eckmann (eds.), *Singular Perturbations and Boundary Layer Theory*, Springer-Verlag, Berlin.

Wanous, D. J. and Sparrow, E. M. (1965). Longitudinal Flow Over a Circular Cylinder with Surface Mass Transfer, *AIAA J.*, vol. 3, pp. 147–149.

Weinbaum, S. and Jiji, L. M. (1977). Singular Perturbation Theory for Melting or Freezing in Finite Domains Initially Not at the Fusion Temperature, *J. Appl. Mech.*, vol. 34, pp. 25–30.

Yang, J. W. (1969). Effect of Uniform Suction on Laminar Film Condensation on a Porous Vertical Wall, ASME Paper No. 69-WA/HT-14.

BIBLIOGRAPHY

PERTURBATION LITERATURE IN HEAT TRANSFER

This bibliography contains a listing of additional literature on perturbation techniques in heat transfer. Classification is made according to different areas of heat transfer. The number in square brackets at the end of each reference indicates the chapter to which the paper belongs. In view of the large body of available literature, only a selection could be included here.

These references should help further consolidate the ideas in the book. Instructors using the book can use these references to formulate additional problems for coursework.

VARIABLE PROPERTY CONDUCTION

Ash, R. L. and Crossman, G. R. (1971). Influence of Temperature Dependent Properties on a Step Heated Semi-Infinite Solid, *J. Heat Transfer*, vol. 93, pp. 250-253. [2]

Cooper, L. Y. (1969). Constant Surface Heating of a Variable Conductivity Half-Space, *Q. J. Appl. Math.*, vol. 27, pp. 173-183. [2]

Cooper, L. Y. (1971). Constant Temperature at the Surface of an Initially Uniform Temperature, Variable Conductivity Half-Space, *J. Heat Transfer*, vol. 93, pp. 55-60. [2]

Seeniraj, V. and Arunachalam, M. (1979). Heat Conduction in a Semi-Infinite Solid with Variable Thermophysical Properties, *Int. J. Heat Mass Transfer*, vol. 22, pp. 1455-1456. [2]

Sodha, M. S., Goyal, I. C., Kaushik, S. C., Tiwari, G. N., Seth, A. K., and Malik, M. A. S. (1979). Periodic Heat Transfer with Temperature Dependent Thermal Conductivity, *Int. J. Heat Mass Transfer*, vol. 22, pp. 777-781. [2]

CONDUCTION IN IRREGULAR DOMAINS

Iabko, I. A. (1973). Unsteady Temperature Field in Plane Bounded Within a Noncircular Contour, *J. Appl. Math. Mech.*, vol. 37, pp. 361-365. [5]

Jiji, L. M. (1974). Singular Perturbation Solutions of Conduction in Irregular Domains, *Q. J. Mech. Appl. Math.*, vol. 27, pp. 45–55. [5]

Williamson, A. S. (1976). The Extraction of Heat from a Bulk Medium through a Thin, Highly-Conducting Disc, *Int. J. Eng. Sci.*, vol. 14, pp. 869–881. [5]

FINS

Aziz, A. (1977). Perturbation Solution for Convective Fin with Internal Heat Generation and Temperature Dependent Thermal Conductivity, *Int. J. Heat Mass Transfer*, vol. 20, pp. 1253–1255. [2]

Aziz, A. and Na, T. Y. (1981). Periodic Heat Transfer in Fins with Variable Thermal Parameters, *Int. J. Heat Mass Transfer*, vol. 24, pp. 1397–1404. [2]

Aziz, A. and Na, T. Y. (1981). Perturbation Analysis for Periodic Heat Transfer in Radiating Fins, *Wärme Stoffübertrag.*, vol. 15, pp. 245–253. [2]

Bilenas, J. A. and Jiji, L. M. (1970). A Perturbation Solution for Fins with Conduction, Convection and Radiation Interaction, *AIAA J.*, vol. 8, pp. 168–169. [2]

CONDUCTION-CONTROLLED PHASE CHANGE

Andrews, J. G. and Atthey, D. R. (1974). Hole Formation in High Power Penetration Welding, CEGB Laboratory Note R/M/N689. [5]

Andrews, J. G. and Atthey, D. R. (1975). Analytical and Numerical Techniques for Ablation Problems, in J. R. Ockendon and W. R. Hodgkins (eds.), *Moving Boundary Problems in Heat Flow and Diffusion*, Clarendon Press, Oxford, pp. 38–53. [5]

Howarth, J. A. and Poots, G. (1976). Inward Solidification with Black Body Radiation at the Boundary, *Mech. Res. Comm.*, vol. 3, pp. 509–514. [5]

Huang, C. L. and Shih, Y. P. (1975). A Perturbation Method for Spherical and Cylindrical Solidification, *Chem. Eng. Sci.*, vol. 30, pp. 897–906. [2]

Huang, C. L. and Shih, Y. P. (1975). Perturbation Solutions of Planar Diffusion-Controlled Moving Boundary Problems, *Int. J. Heat Mass Transfer*, vol. 18, pp. 689–695.

Jiji, L. M. (1970). On the Application of Perturbation to Free Boundary Problems in Radial Systems, *J. Franklin Inst.*, vol. 289, pp. 281–291. [2]

Jiji, L. M. and Weinbaum, S. (1978). Perturbation Solutions for Melting or Freezing in Annular Regions Initially Not at the Fusion, *Int. J. Heat Mass Transfer*, vol. 21, p. 593. [5]

Kuiken, H. K. (1977). Solidification of a Liquid on a Moving Sheet, *Int. J. Heat Mass Transfer*, vol. 20, pp. 309–314. [2]

Lock, G. H. S. (1971). On the Perturbation Solution of the Ice-Water Layer Problem, *Int. J. Heat Mass Transfer*, vol. 14, pp. 642–644. [2]

Lock, G. H. S. and Nyren, R. H. (1971). Analysis of Fully Developed Ice Formation in a Convectively Cooled Circular Tube. *Int. J. Heat Mass Transfer*, vol. 14, pp. 825–834. [2]

Lock, G. H. S., Gunderson, J. R., Quon, D., and Donnelly, J. K. (1969). A Study of One-dimensional Ice Formation with Special Reference to Periodic Growth and Decay, *Int. J. Heat Mass Transfer*, vol. 12, pp. 1343–1352. [2]

Malmuth, N. D. (1976). Temperature Field of a Moving Point Source with Change of State, *Int. J. Heat Mass Transfer*, vol. 19, pp. 349–355. [5]

Rathjen, K. A. and Jiji, L. M. (1970). Transient Heat Transfer in Fins Undergoing Phase Transformation, *Fourth International Heat Transfer Conference*, Section Cu.2.6, pp. 1–11. [2]

Riley, D. S., Smith, F. T., and Poots, G. (1974). The Inward Solidification of Spheres and Circular Cylinders, *Int. J. Heat Mass Transfer*, vol. 17, pp. 1507-1516. [5]

Seeniraj, V. and Bose, T. K. (1982). Planar Solidification of a Warm Flowing Liquid Under Different Boundary Conditions, *Wärme Stoffübertrag.*, vol. 16, pp. 105-111. [2]

Stephan, K. and Holzknecht, B. (1976). Die Asymptotichen Lösungen Für Vorgänge Des Erstarrens, *Int. J. Heat Mass Transfer*, vol. 19, pp. 597-602. [2]

Stewartson, K. and Waechter, R. T. (1976). On Stefan's Problem for Spheres, *Proc. R. Soc. London*, A. 348, pp. 415-426. [5]

Yan, M. M. and Huang, P. N. S. (1979). Perturbation Solutions to Phase Change Problem Subject to Convection and Radiation, *J. Heat Transfer*, vol. 101, pp. 96-100. [2]

CHANNEL FLOW

El-Ariny, A. S. (1978). Steady Pressure Gradient Assisted Plane Couette Flow with Temperature Dependent Viscosity, ASME Paper No. 78-HT-28. [2]

Notter, R. H. and Sleicher, C. A. (1971). A Solution to the Turbulent Graetz Problem by Matched Asymptotic Expansions-II, *Chem. Eng. Sci.*, vol. 26, pp. 559-565. [5]

Richardson, S. M. (1979). Extended Leveque Solutions for Flows of Power Law Fluids in Pipes and Channels, *Int. J. Heat Mass Transfer*, vol. 22, pp. 1417-1423. [2]

Shih, Y. P. and Chen, C. C. (1979). Leveque Series Solutions of Laminar Annular Entrance Heat Transfer, *J. Chin. Inst. Chem. Eng.*, vol. 10, pp. 7-19. [2]

Simon, H. A., Chang, M. H., and Chow, J. C. F. (1977). Heat Transfer in Curved Tubes with Pulsatile, Fully Developed, Laminar Flows, *J. Heat Transfer*, vol. 99, pp. 590-595. [2]

Sleicher, C. A., Notter, R. H., and Crippen, M. D. (1970). A Solution to the Turbulent Graetz Problem by Matched Asymptotic Expansions–I, *Chem. Eng. Sci.*, vol. 25, pp. 845-857. [5]

Turian, R. M. (1965). Viscous Heating in the Cone-and-Plate Viscometer–III, *Chem. Eng. Sci.*, vol. 20, pp. 771-781. [2]

Tyagi, V. P. and Sharma, V. K. (1975). An Analysis of Steady Fully Developed Heat Transfer in Laminar Flow with Viscous Dissipation in a Curved Circular Duct, *Int. J. Heat Mass Transfer*, vol. 18, pp. 69-78. [2]

EXTERNAL BOUNDARY-LAYER FLOW

Curle, S. N. (1980). Calculation of the Axisymmetric Boundary Layer on a Long Thin Cylinder, *Proc. R. Soc. London*, A. 372, pp. 555-564. [6]

Eshghy, S. and Hornbeck, R. W. (1967). Flow and Heat Transfer in the Axisymmetric Boundary Layer Over a Circular Cylinder, *Int. J. Heat Mass Transfer*, vol. 10, pp. 1757-1766. [2, 5]

Fannelop, T. K. (1968). Effect of Streamwise Vortices on Laminar Boundary Layer Flow, *J. Appl. Mech.*, vol. 25, pp. 424-426. [2]

Gupta, B. K. and Kadambi, V. (1971). Effect of Longitudinal Surface Curvature on Heat Transfer, *Wärme Stoffübertrag.*, vol. 4, pp. 9-17. [5]

Kadambi, V. and Gupta, B. K. (1971). Effect of Longitudinal Surface Curvature on Heat Transfer with Dissipation, *Int. J. Heat Mass Transfer*, vol. 14, pp. 1575-1588. [5]

Karniš, J. and Pechoč, V. (1978). The Thermal Laminar Boundary Layer on a Continuous Cylinder, *Int. J. Heat Mass Transfer*, vol. 21, pp. 43-47. [2]

Kuiken, H. K. (1974). The Cooling of a Low-Heat-Resistance Sheet Moving Through a Fluid, *Proc. R. Soc. London*, A. 341, pp. 233-252. [2]

Kuiken, H. K. (1975). The Cooling of a Low-Heat-Resistance Cylinder Moving Through a Fluid, *Proc. R. Soc. London*, A. 346, pp. 23-35. [2]

Kuiken, H. K. (1979). The Cooling of a Low-Heat-Resistance Cylinder by Radiation, *J. Eng. Math.*, vol. 13, pp. 97-106. [5]

Wanous, D. J. and Sparrow, E. M. (1965). Heat Transfer for Flow Longitudinal to a Cylinder with Surface Mass Transfer, *J. Heat Transfer*, vol. 87, pp. 317-319. [2]

NATURAL CONVECTION

Aziz, A. and Na, T. Y. (1982). Improved Perturbation Solutions for Laminar Natural Convection on a Vertical Cylinder, *Wärme Stoffübertrag.*, vol. 16, pp. 83-87. [6]

Chang, K. S., Akins, R. G., and Bankoff, S. G. (1966). Free Convection of a Liquid Metal from a Uniformly Heated Vertical Plate, *I & EC Fundamentals*, vol. 5, pp. 26-37. [2]

Cheng, P. and Lau, K. H. (1974). Steady State Free Convection in an Unconfined Geothermal Reservoir, *J. Geophys. Res.*, vol. 79, pp. 4425-4431. [2]

Chung, P. M. and Anderson, A. D. (1961). Unsteady Laminar Free Convection, *J. Heat Transfer*, vol. 83, pp. 473-478. [2]

Eichhorn, R. (1969). Natural Convection in a Thermally Stratified Fluid, *Prog. Heat Mass Transfer*, vol. 2, pp. 41-53. [2]

Fendell, F. E. (1968). Laminar Natural Convection About an Isothermally Heated Sphere at Small Grashof Number, *J. Fluid Mech.*, vol. 34, pp. 163-176. [5]

Fujii, T. and Uehara, H. (1970). Laminar Natural Convective Heat Transfer from the Outer Surface of a Vertical Cylinder, *Int. J. Heat Mass Transfer*, vol. 13, pp. 607-615. [2]

Kadambi, V. (1969). Singular Perturbations in Free Convection, *Wärme Stoffübertrag.*, vol. 2, pp. 99-104. [5]

Kao, T.-T. (1975). Laminar Free Convective Heat Transfer Response Along a Vertical Flat Plate with Step Jump in Surface Temperature, *Lett. Heat Mass Transfer*, vol. 2, pp. 419-428. [2]

Kelleher, M. (1971). Free Convection from a Vertical Plate with Discontinuous Wall Temperature, *J. Heat Transfer*, vol. 93, pp. 349-356. [2]

Kierkus, W. T. (1968). An Analysis of Laminar Free Convection Flow and Heat Transfer About an Inclined Isothermal Plate, *Int. J. Heat Mass Transfer*, vol. 11, p. 241. [2]

Kuiken, H. K. (1968). Axisymmetric Free Convection Boundary Layer Flow Past Slender Bodies, *Int. J. Heat Mass Transfer*, vol. 11, pp. 1141-1153. [2]

Kuiken, H. K. (1970). Magnetohydrodynamic Free Convection in a Strong Cross Field, *J. Fluid Mech.*, vol. 40, pp. 21-38. [5]

Kuiken, H. K. and Rotem, Z. (1971). Asymptotic Solution for Plume at Very Large and Small Prandtl Numbers, *J. Fluid Mech.*, vol. 45, pp. 585-600. [5]

Lau, K. H. and Cheng, P. (1977). The Effect of Dike Intrusion on Free Convection in Conduction-Dominated Geothermal Reservoirs, *Int. J. Heat Mass Transfer*, vol. 20, pp. 1205-1210. [2]

Mack, L. R. and Bishop, E. H. (1968). Natural Convection Between Horizontal Concentric Cylinders for Low Rayleigh Numbers, *Q. J. Mech. Appl. Math.*, vol. 21, pp. 223-241. [2]

Mack, L. R. and Hardee, H. C. (1968). Natural Convection Between Concentric Spheres at Low Rayleigh Numbers, *Int. J. Heat Mass Transfer*, vol. 11, pp. 387-396. [2]

Sehoenhals, R. J. and Clark, J. A. (1962). Laminar Free Convection Boundary-Layer Perturbations Due to Transverse Wall Vibration, *J. Heat Transfer*, vol. 84, pp. 225-234. [2]

Singh, S. N. (1977). Free Convection from a Sphere in a Slightly Thermally Stratified Fluid, *Int. J. Heat Mass Transfer*, vol. 20, pp. 1155–1160. [5]

Sparrow, E. M. and Gregg, J. L. (1956). Laminar Free Convection Heat Transfer from the Outer Surface of a Vertical Cylinder, *Trans. ASME*, vol. 78, pp. 1823–1829. [2]

Sparrow, E. M. and Gregg, J. L. (1960). Nearly Quasi-Steady Free Convection Heat Transfer in Gases, *J. Heat Transfer*, vol. 82, pp. 258–260. [2]

Yang, K. T. and Jerger, E. N. (1964). First Order Perturbations of Laminar Free Convection Boundary Layers on a Vertical Plate, *J. Heat Transfer*, vol. 86, pp. 107–115. [5]

Yao, L. S. (1980). Analysis of Heat Transfer in Slightly Eccentric Annuli, *J. Heat Transfer*, vol. 102, pp. 279–284. [2]

MIXED CONVECTION

Faris, G. N. and Viskanta, R. (1969). An Analysis of Laminar Combined Forced and Free Convection Heat Transfer in a Horizontal Tube, *Int. J. Heat Mass Transfer*, vol. 12, pp. 1295–1309. [2]

Igbal, M. and Stachiewicz, J. W. (1966). Influence of Tube Orientation on Combined Free and Forced Laminar Convection Heat Transfer, *J. Heat Transfer*, vol. 88, pp. 109–116. [2]

Lin, H.-T. and Shih, Y. P. (1982). Buoyancy Effects on the Forced Convection Along Vertically Moving Surfaces, *Chem. Eng. Comm.*, to be published. [2]

CONVECTION FROM ROTATING SURFACES

Chao, B. T. and Cheema, L. S. (1971). Forced Convection in Wedge Flow with Non-Isothermal Surfaces, *Int. J. Heat Mass Transfer*, vol. 14, pp. 1363–1375. [2]

Sano, T. (1973). A Note on Asymptotic Solutions for Heat Transfer in Laminar Forced Flow Against a Non-Isothermal Rotating Disc, *Wärme Stoffübertrag.*, vol. 2, pp. 108–113. [5]

Tien, C. L. and Tsuji, I. J. (1965). A Theoretical Analysis of Laminar Forced Flow and Heat Transfer About a Rotating Cone, *J. Heat Transfer*, vol. 87, pp. 184–190. [2]

Tsay, S.-Y. and Shih, Y. P. (1977). Perturbation Solutions for the Effect of Prandtl Number on Heat Transfer from a Non-Isothermal Rotating Disc, *Lett. Heat Mass Transfer*, vol. 4, pp. 53–62. [2]

BOILING AND CONDENSATION

Chen, M. M. (1961). An Analytical Study of Laminar Film Condensation, *J. Heat Transfer*, vol. 83, pp. 48–54. [2]

El-Ariny, A. S., Sabbagh, J. A., and Obeid, M. A. (1975). Laminar Film Condensation Heat Transfer in the Presence of Electric and Magnetic Fields, *J. Heat Transfer*, vol. 97, pp. 628–629. [2]

Frankel, N. A. and Bankoff, S. G. (1967). Laminar Film Condensation on a Porous Horizontal Tube with Uniform Suction Velocity, *J. Heat Transfer*, vol. 87, pp. 95–102. [2]

Murty, K. N., Sarma, C. K., and Sarma, P. K. (1975). Laminar Film Condensation on a Vertical Plate-Effect of Magnetic Field, *J. Heat Transfer*, vol. 97, pp. 139–141. [2]

Murty, K. N. and Sarma, P. K. (1977). Laminar Film Condensation–Combined Effect of Magnetic Field and Variable Gravity, ASME Paper No. 77-HT-25. [2]

Nagendra, H. R. (1972). Laminar Film Boiling on Inclined Isothermal Flat Plates, *AIAA J.*, vol. 11, pp. 72–79. [2]

Yih, C.-S. (1980). Flow with Condensation, *Q. Appl. Math.*, pp. 401–409. [2]

CONDUCTION-CONVECTION-RADIATION INTERACTION

Bilenas, J. A. (1974). Transient Perturbation Solution for an Infinite Slab with Radiation, Convection and Conduction Interaction, *Proc. Int. Heat Transfer Conference*, Tokyo. [2]

Dicker, D. and Asnani, M. (1966). A Perturbation Solution for Nonlinear Radiation Heat Transfer, *Proc. Int. Heat Transfer Conference*, Chicago, vol. 5, pp. 164–173. [2]

Tien, C. L. and Abu-Romia, M. M. (1965). Perturbation Solutions in the Differential Analysis of Radiation Interactions with Conduction and Convection, *AIAA J.*, vol. 4, pp. 732–733. [2]

RADIATION IN PARTICIPATING MEDIA

Domoto, G. A. and Wang, W. C. (1974). Radiative Transfer in Homogeneous Nongray Gases with Nonisotropic Particle Scattering, *J. Heat Transfer*, vol. 96, pp. 385–390. [2]

Doura, S. and Howell, J. R. (1977). An Approximate Solution for the Energy Equation with Radiant Participating Media, ASME Paper No. 77-HT-70. [5]

Emanuel, G. (1969). Radiation Energy Transfer from a Small Sphere, *Int. J. Heat Mass Transfer*, vol. 12, pp. 1327–1331. [5]

Jischke, M. C. (1970). Asymptotic Description of Radiating Flow Near a Stagnation Point, *AIAA J.*, vol. 8, pp. 96–101. [5]

Olstad, W. B. (1965). Stagnation-Point Solutions for an Inviscid, Radiating Shock Layer, *1965 Heat Transfer and Fluid Mechanics*, Stanford Univ. Press, pp. 138–156. [4]

INDEX